Certifications of Critical Systems – The CECRIS Experience

RIVER PUBLISHERS SERIES IN INFORMATION SCIENCE AND TECHNOLOGY

Series Editors

K. C. CHEN
National Taiwan University
Taipei, Taiwan

SANDEEP SHUKLA
Virginia Tech, USA
and
Indian Institute of Technology Kanpur, India

Indexing: All books published in this series are submitted to Thomson Reuters Book Citation Index (BkCI), CrossRef and to Google Scholar.

The "River Publishers Series in Information Science and Technology" covers research which ushers the 21st Century into an Internet and multimedia era. Multimedia means the theory and application of filtering, coding, estimating, analyzing, detecting and recognizing, synthesizing, classifying, recording, and reproducing signals by digital and/or analog devices or techniques, while the scope of "signal" includes audio, video, speech, image, musical, multimedia, data/content, geophysical, sonar/radar, bio/medical, sensation, etc. Networking suggests transportation of such multimedia contents among nodes in communication and/or computer networks, to facilitate the ultimate Internet.

Theory, technologies, protocols and standards, applications/services, practice and implementation of wired/wireless networking are all within the scope of this series. Based on network and communication science, we further extend the scope for 21st Century life through the knowledge in robotics, machine learning, embedded systems, cognitive science, pattern recognition, quantum/biological/molecular computation and information processing, biology, ecology, social science and economics, user behaviors and interface, and applications to health and society advance.

Books published in the series include research monographs, edited volumes, handbooks and textbooks. The books provide professionals, researchers, educators, and advanced students in the field with an invaluable insight into the latest research and developments.

Topics covered in the series include, but are by no means restricted to the following:

- Communication/Computer Networking Technologies and Applications
- Queuing Theory
- Optimization
- Operation Research
- Stochastic Processes
- Information Theory
- Multimedia/Speech/Video Processing
- Computation and Information Processing
- Machine Intelligence
- Cognitive Science and Brian Science
- Embedded Systems
- Computer Architectures
- Reconfigurable Computing
- Cyber Security

For a list of other books in this series, www.riverpublishers.com

Certifications of Critical Systems – The CECRIS Experience

Editors

Andrea Bondavalli

Consorzio Interuniversitario Nazionale per l'Informatica (CINI)
and University of Florence
Italy

Francesco Brancati

ResilTech Srl
Italy

Published 2017 by River Publishers

River Publishers
Alsbjergvej 10, 9260 Gistrup, Denmark
www.riverpublishers.com

Distributed exclusively by Routledge

4 Park Square, Milton Park, Abingdon, Oxon OX14 4RN
605 Third Avenue, New York, NY 10158

First published in paperback 2024

Certifications of Critical Systems – The CECRIS Experience / by Andrea Bondavalli, Francesco Brancati.

© The Editor(s) (if applicable) and The Author(s) 2017. This book is published open access.

Open Access
This book is distributed under the terms of the Creative Commons Attribution-Non-Commercial 4.0 International License, CC-BY-NC 4.0) (http://creativecommons.org/licenses/by/4.0/), which permits use, duplication, adaptation, distribution and reproduction in any medium or format, as long as you give appropriate credit to the original author(s) and the source, a link is provided to the Creative Commons license and any changes made are indicated. The images or other third party material in this book are included in the work's Creative Commons license, unless indicated otherwise in the credit line; if such material is not included in the work's Creative Commons license and the respective action is not permitted by statutory regulation, users will need to obtain permission from the license holder to duplicate, adapt, or reproduce the material.

The use of general descriptive names, registered names, trademarks, service marks, etc. in this publication does not imply, even in the absence of a specific statement, that such names are exempt from the relevant protective laws and regulations and therefore free for general use.
The publisher, the authors and the editors are safe to assume that the advice and information in this book are believed to be true and accurate at the date of publication. Neither the publisher nor the authors or the editors give a warranty, express or implied, with respect to the material contained herein or for any errors or omissions that may have been made.

Routledge is an imprint of the Taylor & Francis Group, an informa business

Publisher's Note
The publisher has gone to great lengths to ensure the quality of this reprint but points out that some imperfections in the original copies may be apparent.

While every effort is made to provide dependable information, the publisher, authors, and editors cannot be held responsible for any errors or omissions.

ISBN: 978-87-93519-56-5 (hbk)
ISBN: 978-87-7004-417-2 (pbk)
ISBN: 978-1-003-33748-5 (ebk)

DOI: 10.1201/9781003337485

Contents

Preface

The rapid spread of critical systems raises new challenges from multiple aspects. The functionality embedded into critical systems is a major driver of efficient and economic operation of a variety of societal services ranging from traffic control to health care, but at the same time, the vulnerability of the society to malfunctioning equipment reaches a critical level both in the terms of risks to the human life and huge economic impacts. The rapid development of underlying technologies implies a huge challenge to this industry which followed for decades a safety driven conservative approach. This way, a uniform approach to the development, validation and verification is an important factor in the Europe wide integration of services as emphasized for instance by the creation of the ARTEMIS European Technology Platform on the side of technology. On the human skill side, the dissemination of the best industrial practices and appropriate training is a key enabling factor for this unification process.

All over Europe there is a significant lack of skilled workforce related to critical embedded systems.

Traditional V&V methods frequently exceed effort needed for the core development time, and while the "soft" IT industry rapidly turns to system integration based on the reuse of high volume hardware and software components, for safety related applications this will still evolve.

All this poses serious difficulties to companies, which are on one hand constrained to meet predefined quality goals, whereas, on the other hand, are required to deliver systems at acceptable cost and time to market. Large companies mainly follow a brute-force approach by focused large volume investment into tooling and in-house training, but even high-tech SMEs are highly vulnerable to the new challenges.

Looking at the field of the Verification and Validation one of the most challenging goals is the definition of methods, strategies and tools able to validate a system adequately, while simultaneously keeping the cost and delivery time reasonably low. It is not easily possible to establish a proper balance between achievable quality with a particular technique (in terms of RAMS attributes) and the costs required for achieving such quality. The situation is even worse in the case of integration of existing SW in a safety critical system to be certified, since, assessing products which encompass COTS software is a challenge although modern standards consider this possibility. An additional concern is the usage of recently adopted methods for SW development like model based ones, since the certification of systems using software developed with these supports is at the limit of the applicability of the existing standards, and only the most recent ones are aligned with these 'modern' methods.

This book documents the main insights on Cost Effective Verification and Validation processes that we gained during our work in the European Research Project CECRIS (acronym for *Certification of Critical Systems*). The objective of this research was to tackle the challenges of certification by focusing on those aspects that turn out to be more difficult and or important for current and future critical systems industry: the effective use of methodologies, processes and tools.

The CECRIS project took a step forward in the growing field of development, verification and validation and certification of critical systems. It focused on the more difficult/important points of (safety, efficiency, business) of critical system development, verification and validation and certification process. The scientific objectives of the project were to study both the scientific and industrial state of the art methodologies for system development and the impact of their usage on the verification and validation and certification of critical systems. Moreover the project aimed at developing strategies and techniques supported by automatic or semi-automatic tools and methods for these types of activities, whose cost-quality achievements are well-predictable in order to tie costs of application of techniques to the RAMS attributes level achieved by the product being tested. The project set guidelines to support engineers during the planning of the verification & validation phases.

The Project Consortium was composed by three academic partners and three companies:

1. CINI-Consorzio Interuniversitario Nazionale per l'Informatica
2. Resiltech S.r.l.
3. Universidade de Coimbra
4. Budapesti Muszaki es Gazdasagtudomanyi Egyetem
5. Prolan Iranyitastechnikai Zartkoruen Mukodo Reszvenytarsasag
6. CRITICAL Software SA

The CECRIS project has given to the partners the opportunity of sharing their industrial-academic expertise and experiences and to develop fruitful collaborations and research products. Through the 'Transfer of Knowledge' activities, industrial partners have had the opportunity to better know, evaluate and apply new research methods, while the academic partners could get from industry valuable feedback, better understanding the industrial problems and needs.

Several synergies that have been established during the secondments, are now in place beyond the project termination for exploiting further potential strategic research activities. Moreover, the collaborations for the maintenance and improvement of the project tools developed during CECRIS will last for years, since these tools support the overall V&V process and reduce the certification costs of safety-critical systems.

It is the objective of this book to collect the main project results in terms of methodologies and processes and to propose them in a single edited book.

The first part of the book is related to certification processes. Chapter one presents an easy-to-use framework and a supporting methodology to perform a rapid gap analysis on the usage of standards for safety-critical software, being them new ones to be introduced or standards already applied. In other words, the framework can be applied to reason in terms of "changing standard" or in terms of "introducing a new standard". The ultimate objective is to discover with limited effort how far a company is from acquiring sufficient the necessary and sufficient level of knowledge to apply a specific standard. Our approach is based on the concept of rating the knowledge available: it starts from an understanding of the expertise of a company, and it rates the improvements, in terms of training, needed to reach an adequate level of confidence with the techniques and processes required in the standard. Our approach can be applied to an entire standard, a part of it, or to individual techniques and tools. Thus, our framework offers the possibility to depict the status of the knowledge available in the company, which may offer valuable insights on the areas that are mostly covered, and where potential improvements are possible. The approach can indicate the introduction time, which estimates the overall training time required to introduce a new standard.

The second part of the book focuses on model-driven methodologies. For a company being competitive on the market, following technologies and being updated with new trends and practices is essential. In safety-critical domains, the introduction of new practices and methodologies is slower than in other engineering fields, since safety standards and long established practices tend to defer the adoption of new emerging technologies, until assessments and time reveal them mature and safe enough. Slow introduction of new methods is especially characterizing the railway domain where the lifespan of products could easily reach decades or even a century. Now it is long time that Model-Driven Engineering techniques and tools have been proposed, but their maturity – especially for safety-critical systems – is still debated. Some recent surveys investigated the adoption of MDE methodologies and technologies in practice. They revealed the increasing adoption of MDE in industry. The technology is attractive for the development of critical systems, since it can speed up the activities of Verification and Validation (V&V), and it enables the early verification of systems, through techniques such as model reviews, guideline checkers, rapid control prototyping and model- and software-in-the-loop Tests. These techniques shift the cost of development from the phases of V&V to the ones of requirement analysis and design, thus leading to benefits in terms of residual errors. Companies not performing model-in-the-loop testing find almost 30% more errors during module test. Chapter two reports the results of a twelve months industrial-academic partnership for the transfer of knowledge of MDE techniques from the academy to one of the company involved in the project, with the goal of assessing their level of maturity for industrial adoption. During this activity, it emerged the lack of well-defined processes for the development of a CENELEC SIL-4 safety critical signaling system that was suited for the real industrial needs.

In Chapter three focuses on the issues related to the lack of expertise in CS/OO/SysML formalisms that often lead to the need of a lot of training and support to use the modeling tools. Ideally, designers should spend all their effort on modeling and nothing else. However, existing modeling tools have lot of issues related to installation and plug-ins. The use of Google Blockly was envisaged for modeling and simulation of systems. Blockly is a visual programming library, used to model/program using interlocked blocks. Each of the blocks also supports traditional input widgets such as labels, images, textbox, checkbox, combo box, etc. It can be configured in such a way that only compatible blocks can be connected together (i.e. can be made "valid by design"). Blockly supports code and XML generation, and requires only a modern web browser which can be run on any device or operating system. However, Blockly was not readily usable for modeling using SysML/UML like formalisms. A lot of changes and customizations were made in Blockly to make it more suitable for such type of modeling.

The Third part of this book composed of Chapters four, five, six and seven, deals with V&V and quality processes.

Chapter four presents a process for finding and tackling the main root causes that affect critical systems quality. Following standards and applying good engineering practices during software development is not enough to guarantee defects free software, thus additional processes, such as Independent Software Verification and Validation (ISVV), are required in critical projects. The objective of ISVV is to provide complementary and independent assessments of the software artifacts in order to find residual defects and allow their correction in a timely manner. Independence is the most important concept of ISVV and it has been referred to and used in safety-critical domains such as civil aviation (DO-178B), railway signaling systems (CENELEC), and space missions (European Cooperation for Space Standardization – ECSS). However, such systems are still far from being perfect and it is common to hear about software bugs in aeronautics, train accidents caused by software problems, satellite systems that need to be patched after launch, and so on. This chapter presents an analysis on trends, common (and uncommon) problems and their causes, and looks at the general picture of critical defects within the software development lifecycle of space systems, considering a dataset of 1070 defects. The results are intended to help engineers in tackling the problems starting from the most frequent ones, instead of dealing with them one by one, as is traditionally done in industry nowadays. In practice, this work brings light to the main root causes of issues in space projects, which were identified, based on the defects classification and on relevant expert knowledge about those defects and about the software development process, contributing towards proposing improvements to the processes, methodologies, tools, standards and industry culture.

Chapter 5 describes a framework for automation of hazard log management on large critical projects. A hazard is any situation that could cause harm to the system or lives. Hazards depend on the system and its environment, and the probability of the hazard to cause harm is known as risk. Hazards are analyzed by identifying their causes and the possible negative consequences that might ensue. This chapter describes a modular and extensible way to specify rules for checks locally at the stake-holder side, as well as while combining data from various parties to form the hazard log (HL). The HZ-LOG automatization tool simplifies the process of hazard data collection on large projects to form the hazard log while ensuring data consistency and correctness. The data provided by all parties are collected using a template containing scripts to check for mistakes/errors based on internal standards of the company in charge of the hazard management. The collected data is then subjected to merging in DOORS, which also contain scripts to check and import data to form the hazard log.

Chapter 6 instead deals with cost estimation for independent systems verification and validation. Validation, verification and especially certification are skill and effort demanding activities which are typically performed in an independent way by specialized small and medium enterprises. Prediction of the work needed to accomplish them is crucial for the management of such projects, which is by its very nature heavily depending on the implementation of the V&V process and its support. Process management widely uses cost estimators in planning of software development projects for resource allocation. Cost estimators use the scoring of a set of cost influencing factors, as input. They use extrapolation functions calibrated previously on measures extracted from a set of representative historical project records. These predictors do not provide reliable measures for the separate phases of V&V and certification in safety critical projects. The current chapter summarizes the main use cases and results of an activity focusing on these particular phases.

Chapter 7 addresses lightweight formal analysis of requirements which are the core items of the design (and Validation) workflow of safety critical systems. Accordingly, their completeness, compliance with the standards and understandability is a dominant factor in the subsequent steps. Requirements review is a special kind of Independent Software/Systems Verification and Validation (ISVV). The chapter presents methodologies to use lightweight formal methods supporting experts in a peer review based ISVV.

Part four of this book, composed of chapters eight and nine, deals with particular phases of V&V processes known as FMEA & FMECA.

Chapter 8 describes STECA which stands for "Security Threats, Effects and Criticality Analysis" and its application to a Smart Grids scenario. The STECA approach is meant to perform security assessment and the chapter explains the process proposed to identify vulnerabilities, their related threats, a risk assessment approach and finally a path to identify appropriate countermeasures. This process is based on the same principles used for the FMEA/FMECA process, widely used for safety critical analysis and highly regarded by the majority of international standards. STECA starts from a vulnerability point of view and moves on towards threat analysis and criticality assessment. Following the guidelines defined, the approach is then instantiated on a Smart Grid use case, resulting in a set of precise guidelines and a systematic way to perform security assessment including vulnerability evaluation and attack impact analysis.

Chapter 9 describes a composable framework support for Software-FMEA through Model Execution. Performing Failure Mode and Effects Analysis (FMEA) during software architecture design is becoming a basic requirement in an increasing number of domains. However, due to the lack of standardized early design-phase model execution, classic Software-FMEA (SW-FMEA) approaches carry significant risks and are human effort-intensive even in processes that use Model-Driven Engineering.

From a dependability-critical development process point of view, FMEA should be performed in the early phases of system design; for software, this usually translates to the architecture design phase. Additionally, for some domains, standards prescribe the safety analysis of the software architecture – as is the case e.g. with ISO 26262 in the automotive domain. Significant risk is introduced by the fact that the error propagation assumptions usually made at this stage have to hold for the final system – otherwise the constructed hazard mitigation arguments will not hold. This chapter addresses SFMEA based on a new standard for UML 2 modeling language. Throughout the chapter, the reader will be introduced to i) advances in standardized model execution semantics, ii) the outline of a composable framework built on top of executable software architecture models to help SW-FMEA, iii) a realization of such a framework applied on a case study from the railway domain.

The last part of this book, Part five, contains contributions developed in CECRIS related to Robustness and Fault injection and is composed of 3 chapters.

Chapter 10 describes a monitoring and testing framework for critical off-the-shelf applications and services. One of the biggest verification and validation challenges is the definition of approaches and tools to support systems assessment while minimizing costs and delivery time. Such tools reduce the time and cost of assessing Off-The-Shelf (OTS) software components that must undergo proper certification or approval processes to be used in critical scenarios. In the case of testing, due to the particularities of components, developers often build ad-hoc and poorly-reusable testing tools, which results in increased time and costs. This chapter introduces a framework for testing and monitoring of critical OTS applications and services. The framework includes i) a box instrumented for monitoring OS and application level variables, ii) a toolset for testing the target components and iii) tools for data storing, retrieval and analysis. The chapter presents an implementation of the framework that allows applying, in a cost-effective fashion, functional testing, robustness testing and penetration testing to web services. Finally, the framework usability and utility is demonstrated based on two different case studies that also show its flexibility.

Chapter 11 is about the validation of a safety critical railway application using fault injection. This chapter will summarize the fault injection experiments performed with the ProSigma system. It will include a detailed description of the system, fault injection test goals, description of the fault injection tool, the results of the FI tests, etc.

Chapter 12 is concerned with robustness of complex Critical Systems. Systems are nowadays being deployed also as services or web applications, and are being used to provide enterprise-level business-critical operations. These systems are supported by complex middleware, which often links different systems, and where a failure can bring in disastrous consequences for both clients and service providers. In this chapter we present a toolset that can be used to evaluate the robustness of a given system, under the following two different perspectives: i) executing robustness tests against the service's external interface (e.g., the interface with business clients) and also inner interfaces (e.g., the application-database interface); ii) emulating the presence of source code defects, on the service middleware, to understand how the presence of a defect can affect the robustness of the overall system. The toolset has been demonstrated on a set of web services, an Enterprise Resource Planning web application, and on the popular Apache HTTP server. Results show that the toolset can be easily used to disclose critical problems in web applications and to support middleware, helping developers in building and validating more reliable services.

Although the chapters of the book are arranged in a logical order, an effort has been made to keep each chapter self-contained. This book can be used for supplemental reading for advanced teaching on Critical systems validation and verification methodologies.

Andrea Bondavalli

Francesco Brancati

List of Contributors

Alexandre Esper, *CRITICAL Software S.A., Coimbra, Portugal*

András Pataricza, *Dept. of Measurement and Information Systems, Budapest University of Technology and Economics, Budapest, Hungary*

András Zentai, *Prolan Process Control Co., Szentendrei út 1–3, H-2011 Budakalász, Hungary*

Andrea Bondavalli, *1) Department of Mathematics and Informatics, University of Florence, Florence, Italy*
2) CINI-Consorzio Interuniversitario Nazionale per l'Informatica-University of Florence, Florence, Italy

Andrea Ceccarelli, *1) Department of Mathematics and Informatics, University of Florence, Florence, Italy*
2) CINI-Consorzio Interuniversitario Nazionale per l'Informatica-University of Florence, Florence, Italy

Arun Babu Puthuparambil, *Robert Bosch Center for Cyber Physical Systems, Indian Institute of Science, Bangalore, India*

Cristiana Areias, *1) CISUC, Department of Informatics Engineering, University of Coimbra, Portugal*
2) ISEC – Coimbra Institute of Engineering, Polytechnic Institute of Coimbra, Portugal

Fabio Scippacercola, *1) DIETI, Università degli Studi di Napoli Federico II, Via Claudio 21, 80125 Napoli, Italy*
2) CINI-Consorzio Interuniversitario Nazionale per l'Informatica, Italy

Francesco Brancati, *Resiltech s.r.l., Pontedera (PI), Italy*

Francesco Rossi, *Resiltech s.r.l., Pontedera (PI), Italy*

Francisco Moreira, *CRITICAL Software S.A., Coimbra, Portugal*

Gonçalo Pereira, *CISUC, Department of Informatics Engineering, University of Coimbra, Portugal*

Henrique Madeira, *CISUC, Department of Informatics Engineering, University of Coimbra, Portugal*

Imre Kocsis, *Dept. of Measurement and Information Systems, Budapest University of Technology and Economics, Budapest, Hungary*

Ivano Irrera, *CISUC, Department of Informatics Engineering, University of Coimbra, Portugal*

João Carlos Cunha, *1) CISUC, Department of Informatics Engineering, University of Coimbra, Portugal*
2) ISEC – Coimbra Institute of Engineering, Polytechnic Institute of Coimbra, Portugal

Jorge Bernardino, *1) CISUC, Department of Informatics Engineering, University of Coimbra, Portugal*
2) ISEC – Coimbra Institute of Engineering, Polytechnic Institute of Coimbra, Portugal

László Gönczy, *Dept. of Measurement and Information Systems, Budapest University of Technology and Economics, Budapest, Hungary*

Leonardo Montecchi, *1) Department of Mathematics and Informatics, University of Florence, Florence, Italy*
2) CINI-Consorzio Interuniversitario Nazionale per l'Informatica-University of Florence, Florence, Italy

Lorenzo Vinerbi, *Resiltech s.r.l., Pontedera (PI), Italy*

Marco Vieira, *CISUC, Department of Informatics Engineering, University of Coimbra, Portugal*

Mario Rui Baptista, *CRITICAL Software S.A., Coimbra, Portugal*

Nicola Nostro, *Resiltech s.r.l., Pontedera (PI), Italy*

Nuno Antunes, *CISUC, Department of Informatics Engineering, University of Coimbra, Portugal*

Nuno Laranjeiro, *CISUC, Department of Informatics Engineering, University of Coimbra, Portugal*

Nuno Silva, *CRITICAL Software S.A., Coimbra, Portugal*

Raul Barbosa, *CISUC, Department of Informatics Engineering, University of Coimbra, Portugal*

Rosaria Esposito, *Resiltech s.r.l., Pontedera (PI), Italy*

Seyma Nur Soydemir, *CISUC, Department of Informatics Engineering, University of Coimbra, Portugal*

Stefano Russo, *1) DIETI, Università degli Studi di Napoli Federico II, Via Claudio 21, 80125 Napoli, Italy*
2) CINI-Consorzio Interuniversitario Nazionale per l'Informatica, Italy

Tommaso Zoppi, *1) Department of Mathematics and Informatics, University of Florence, Florence, Italy*
2) CINI-Consorzio Interuniversitario Nazionale per l'Informatica-University of Florence, Florence, Italy

Valentina Bonfiglio, *Resiltech s.r.l., Pontedera (PI), Italy*

Vince Molnár, *Dept. of Measurement and Information Systems, Budapest University of Technology and Economics, Budapest, Hungary*

List of Figures

List of Tables

List of Abbreviations

A/D	Analog/Digital
ALARP	As low as reasonably practicable
Alf	Action Language for Foundational UML
ASILs	Automotive Software Integrity Levels
ASPICE	Automotive SPICE
BB-PIT	Black Box Platform Independent Test Model
BB-PST	Black Box Platform Specific Test Model
BDD	Block Definition Diagram
BI	Business Intelligence
CAN	Controller Area Network
CE	Cost Estimator
CENELEC	Comité européen de normalisation en électronique et en électrotechnique
CIM	Computation Independent Model
CIT	Computation Independent Test Model
CIV	Computation Independent Viewpoint
CMMI	Capability Maturity Model Integration
COCOMO	Constructive Cost Model
COTS	Commercial Off-The-Shelf
CPU	Central Processing Unit
CS	Critical System
CSP	Constraint Satisfaction Problem
csp(FD)	finite-domain CSP
CTC	Central Traffic Control
Dako	Andras, need your help here
DB	Database
DI	Digital Input
DMI	*Driver Machine Interface*
DOORS	Dynamic Object Oriented Requirements System
DSL	Domain-Specific Language

Eclipse RMF	Eclipse Requirement Management Framework
ECSS	European Cooperation for Space Standardization
EER	Enhanced Entity–Relationship
EN	Européen Norme
ERTMS	European Rail Traffic Management System
ESA	European Space Agency
ETCS	European Train Control System
ETH	CAN to UDP protocol converter
FDIR	Fault Detection, Isolation and Recovery
FI	Fault Injection
FIR	Fault Injection Runs
FIT	Fault Injection Tool
FMEA	Failure Modes and Effects Analysis
FMECA	Failure Modes, Effects and Criticality Analysis
FTA	Fault Tree Analysis
FW	Firmware
GA	Generic Application
GB-PIT	Grey Box Platform Independent Test Model
GP	Generic Product
GR	Golden Runs
GSM	Global System for Mobile communications
HA	Hazard analysis
HAN	Home Area Network
HB	HeartBeat signal
HIL	Hardware-in-the-loop
HL	Hazard log
HMI	Human-Machine Interface
HSIA	HW/SW interaction analysis
HW	Hardware
HZ	Hazard
IBD	Internal Block Diagram
ICT	Information and Communication Technology
IDEF	Integration DEFinition
IEC	International Electrotechnical Commission
IP	Internet Protocol
IS	Interlocking System
ISO	International Organization for Standardization
ISVV	Independent Software Verification and Validation

ISVV	Independent Software/Systems Verification and Validation
JIF	Relay Interface
JTAG	Join Test Action Group
KLOC	Thousands of lines of code
KPI	Key Performance Indicator
LI	Logic and Input
M2M	Model-to-Model Transformation
M2T	Model-to-Text Transformation
MBE	Model-Based Engineering
MBSE	Model-Based System Engineering
MDA	Model-Driven Architecture
MDD	Model-Driven Development
MDE	Model-Driven Engineering
MDT	Model-Driven Testing
MIL	Model-in-the-loop
MoC	Models of computation
MT	Mitigation
NIST	National Institute of Standards and Technology
OBU	On-board Unit
OCD	On-Chip Debugger
ODC	Orthogonal Defect Classification
OMG	Object Management Group
OS	Operating System
OWL	Web Ontology Language
OXF	Object Execution Framework
PA	Product Assurance
PAR	Parameter Module
PB	Prolan Block
PHA	Preliminary Hazard Analysis
PIM	Platform Independent Model
PIT	Platform Independent Test Model
PIV	Platform Independent Viewpoint
PM	Prolan Monitor
PSDK	Prosigma Diagnostic Center
PSM	Platform Specific Model
PST	Platform Specific Test Model
PSU	Power Supply Unit
PSV	Platform Specific Viewpoint
PTD	ProSigma generic application

QA	Quality Assurance
RAM	Random Access Memory
RAMS	Reliability, Availability, Maintainability, and Safety
RBC	Radio Block Control
RCA	Root Cause Analysis
RDF	Resource Description Framework
ReqIF	Requirements Interchange Format
RID	Review Identified Discrepancy
RODIN	Rigorous Open Development Environment for Complex Systems
ROI	Return on Investment
RPI	UDP to X25 over IP protocol converter
SA	Specific Application
SAM	Specific Application Module
SCAMPI	Standard CMMI Appraisal Method for Process Improvement
SDLC	Software Development Life Cycle
SDP	Software Development Process
SHA	System hazard analysis
SIL	Safety Integrity Level
SME	Small and medium-sized enterprise
SPICE	Software Process Improvement and Capability Determination
SSHA	Subsystem hazard analysis
SST	Safety Signal Transmitter
STECA	Security Threats, Effects and Criticality Analysis
SUT	System Under Test
SVF	Software Validation Facility
SW	Software
SW-FMEA	Software Failure Modes and Effects Analysis
SXF	Simple Execution Framework
SysML	Systems Modeling Language
OMG	
TC	Telecommand
TIU	*Train Interface Unit*
TM	Telemetry
TMR	Triple Modular Redundancy
UDP	User Datagram Protocol
UML	Unified Modeling Language

USB	Universal Serial Bus
UTP	UML Testing Profile
V&V	Verification and Validation
W3C	World Wide Web Consortium
WB-PST	White Box Platform Specific Test Model
X25	ITU-T X.25 Protocol

1

A Framework to Identify Companies Gaps When Introducing New Standards for Safety-Critical Software

Andrea Ceccarelli[1,2] and Nuno Silva[3]

[1]Department of Mathematics and Informatics, University of Florence, Florence, Italy
[2]CINI-Consorzio Interuniversitario Nazionale per l'Informatica-University of Florence, Florence, Italy
[3]CRITICAL Software S.A., Coimbra, Portugal

1.1 Introduction

Companies working in safety-critical domains as the avionics or space have mandatorily to comply with standards, regulating the lifecycle of the system development, the techniques to be adopted and requirements to be fulfilled in different lifecycle phases. Consequently, a company that develops systems or products in compliance with a standard need skill to use the recommended techniques, often with the support of tools developed within the company or from third parties.

The variety of Information and Communications Technology (ICT) world and applicable domains nowadays imply that several standards for safety-critical systems exist, applied mandatorily and regulating the development and operation of critical systems. As examples, the DO-178B/C [1, 2], DO-254 [3] are the mandatory international standards for the avionics domain; the CENELEC EN 50126 [4], 50128 [5] and 50129 [6] are the mandatory standards for the European railway domain; the ECSS [7] is the set of standards for the space domain in Europe.

When changing domain, a company needs to apply different standards and can encounter several connected issues, such as different: (i) definitions; (ii) level of expectations; (iii) level of details of the required tasks;

(iv) maturity level of processes, techniques, tools, customers, etc.; (v) requirements for tool qualification.

A company wanting to adopt a standard, e.g., as the consequence to the decision to enter a new market, must necessarily (i) gain the skills, techniques and tools necessary to appropriately operate in compliance with the standard, (ii) have a different mindset, and (iii) acquire the necessary expertise. The question that is naturally raised is related to the effort, both in time and cost, of introduction of a standard in a company. Such an effort can be considerable, if the company never worked with similar standards or domains.

1.1.1 Contribution

We present *an easy-to-use framework and a supporting methodology to perform a rapid gap analysis on the usage of standards for safety-critical software*, being them new ones to be introduced or already applied. In other words, the framework can be applied to reason in terms of "changing standard" or in terms of "introducing a new standard". The ultimate objective is to discover with limited effort how far a company is from acquiring a level of knowledge sufficient to apply a specific standard. Our approach is based on the concept of rating the knowledge available: it starts from an understanding of the expertise of a company, and it rates the improvements, in terms of training, needed to reach an adequate level of confidence with the techniques and processes required in the standard. Our approach can be applied to a whole standard, a part of it, or to individual techniques and tools. Thus, our framework offers the possibility to depict the status of the knowledge available in the company, which may offer valuable insights on the areas that are mostly covered, and where potential improvements are possible. The approach can indicate the introduction time, which estimates the overall training time required to introduce a new standard.

In the case study, the framework and the supporting methodology are applied to investigate the verification and validation phases of the DO-178B standard in the company *CRITICAL Software S.A.*

We note that our framework cannot be dissociated from the personnel operating in the company: in fact, the personnel are actually holding the background knowledge and are in charge of acquiring new knowledge. Consequently, the identification of the personnel in the company and their role, together with an investigation of their skills, is part of our approach and connected to the outcome of the analysis.

A relevant note is that we specifically target *software companies*, *prescriptive standards for software*, and the *safety-critical domains*. Although the framework may also be applicable to other kinds of companies, standards (e.g., goal-based standards opposed to prescriptive ones [8]), and domains, we explicitly remark that our investigations, use case and claims of validity are exclusively related to the above targets. A preliminary version of the framework and methodology appeared in [9].

The rest of this chapter is organized as follows. Section 1.2 presents the state of the art. Section 1.3 illustrates the framework and the methodology. Section 1.4 presents the structure of the dataset used, and how to populate it. Section 1.5 presents the metrics for the qualitative and quantitative evaluation of gaps. Section 1.6 presents the case study, Section 1.7 discusses relevant arguments to exercise the framework, and Section 1.8 concludes the Chapter.

1.2 State of the Art on Gap Analysis in the ICT World

Gap analysis is a renowned concept that finds application in several fields since many years; significant examples are in the fields of civil engineering [10], biology [11], economics [12] and ICT [13–18].

In ICT, gap analysis is usually defined as the study of the differences between two information systems or applications, often for the purpose of determining how to get from one state to a new state. A gap can be presented as the space between where we are and where we want to be; gap analysis is undertaken as a mean of bridging that space. We report on most relevant examples of gap analysis for safety-critical systems.

Gap analysis is part of the Software Process Improvement and Capability Determination (SPICE, now an ISO/IEC standards set [14]) to afford the process capability level evaluations of suppliers. SPICE can result useful to select the cheapest supplier amongst those with adequate qualification, or to identify gaps between the current capability of the supplier and the level required by a potential customer. Similarly, the Automotive SPICE (ASPICE, [195]) starts from SPICE but is specific to the automotive industry. Furthermore, the Capability Maturity Model Integration (CMMI, [13]) includes the Standard CMMI Appraisal Method for Process Improvement (SCAMPI, [13]) that is aimed to appraise organizations capability maturity; the SCAMPI approach can result in a capability level profile, or also in benchmarking against other organizations. However, evaluating performance lies out of its scope [20]. CMMI compliance is not a guarantee of good performance *per se*, i.e., there is high variance in performance results within a maturity level [20]. According to [21, 22], in general, these structured processes are widely applicable for large organizations, while their suitability is more arguable for smaller ones. For both large and small organizations, main concerns are the often elevated costs, the highly complex recommendations, and the improvement projects which involve a large investment in terms of money, time, resources and long time to benefit.

There are several other examples of gap analysis in the ICT. The Integration DEFinition (IDEF, [15]) is a group of methods used to create a model of a system, analyze the model, create a model of a desired version of the system, and aid in the transition from one to the other. [16] defines an index for measuring and analyzing the divide among countries in the area of ICT infrastructure and access [17] develops a Skills Gap Analysis study to respond to immediate inquiries for information on the needs for ICT skills covering the local, regional, and global markets [18] explores the determinants of cross-country disparities in personalcomputer and Internet penetration, relating technology penetration rates with income, human capital, the youth dependency ratio, telephone density, legal quality, and banking sector development.

Other related approaches can be identified in methods for evaluating the cost of software development projects (e.g., COCOMO [23]), as well as system engineering costs (e.g., COSYSMO [24]). Additionally, there have been efforts in building frameworks to guide and support the design, assessment and certification process, for example [25, 26].

Summarizing, overall a vast literature exists on gap analysis, introduction time, and compliance to safety-critical standards. To the authors' knowledge, till today there are no publicly-available gap analysis for software safety standards that are easy-to-use, easy-to-maintain, and that allows understanding, with limited investment, the effort required to become confident with a standard. Companies can benefit from our solution to evaluate their expertise with a standard, measure how difficult it would be to introduce it, and define an appropriate plan for such a standard.

1.3 Overview of the Framework and Methodology

The framework and the related methodology are presented in this Section. They can be realized and executed with the support of a database and tools for drafting questionnaires and data analysis tools. In fact, the whole methodology was implemented and exercised using as supporting instruments a MySQL database to store data, MySQL Workbench to ease database management, the Google Docs suite to make questionnaire and reports, and a few Java classes for data extraction and elaboration, and to implement the decision tree of Section 1.5.

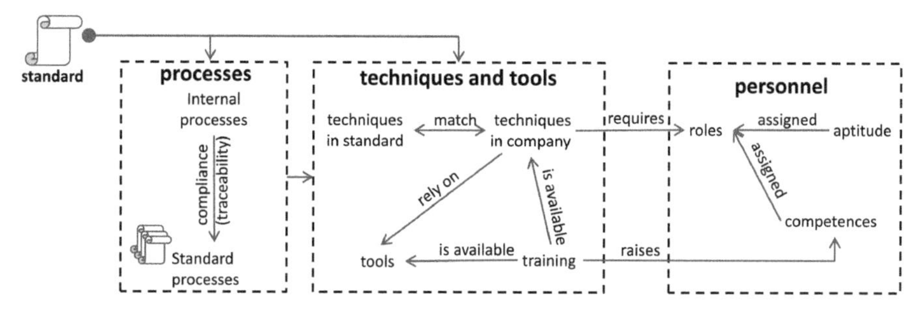

Figure 1.1 Overall view of the gap analysis framework.

In the following chapter, to include examples and to guide our case study, we refer as background knowledge to [27] that classifies the main items, techniques, and processes of aerospace software standards.

1.3.1 The Framework

We present the overall framework with the support of Figure 1.1. It is structured in three main blocks: *Processes*, *Techniques, and Tools*, and *Personnel*. The input to the first two blocks is the *standard* under analysis.

1.3.1.1 Processes

This block is devoted to the identification and matching of the processes. It contains *internal processes* and *standard processes*. Internal processes are defined and applied in a company e.g., internal quality management systems, or internal processes that are required for having certifications like ISO 9001 [28] or CMMI. *Standard processes* are instead the processes or requirements defined in standards; examples at a macro level are design, development, verification, validation, or integration processes.

For each standard, a corresponding traceability matrix must be created and populated; it checks that *internal processes* are compliant to *standard processes*. One or more internal processes should be matched to each process of each individual standard. If the matching is not complete, there may be the necessity to review internal processes; otherwise the applicability of the standard may be compromised.

Although solutions to automate these checks exist [29, 30], we believe that a visual inspection of the standard is sufficient to identify major inconsistencies. This claim is supported by the typically structured descriptions of internal processes and standard processes.

The identification and matching of such processes are inputs to the block Techniques and Tools.

1.3.1.2 Techniques and tools

Both standard processes and internal processes typically list recommended or mandatory techniques.

A whole list of techniques in the standard (*techniques in standard*) and techniques available in the company (*techniques in company*) is required. The list of the techniques in standard needs to be compiled for each standard; the list of techniques in company needs to be compiled only once, and updated when a new technique is learnt.

A traceability matrix can match techniques in company and techniques in standard, to identify the correspondence between the two or possible mismatches. For example, a technique discussed in a standard that has no correspondence among the techniques available in the company know-how. One or more *techniques in company* may be matched to each *technique in standard*. Techniques in standard and techniques in company are also matched to, respectively, standard processes and internal process.

Tools are connected to the techniques in the company, because they can support their execution (occasionally tools can support the whole process [25, 26], although this possibility is not represented in Figure 1.1). Similarly, *training* materials (e.g., slides from courses or tutorials), whenever available, are enlisted and mapped to the company tools and techniques. Noteworthy, techniques or tools not explicitly mentioned in internal processes may be available in the company and useful to support the execution of such internal processes: in this case, it is required to add such techniques or tools and create the appropriate connections to the internal processes.

It is fundamental to understand the confidence in using a technique or a tool; an option is to acquire this information through a questionnaire, as we will discuss in later sections of this chapter. Obviously, this has not to be done on individual basis to rate the single worker, but as a collective exercise between expert workers.

1.3.1.3 Personnel

The personnel are actually holding the background knowledge of the company and are in charge of acquiring new knowledge. The block *personnel* relate the company's personnel to the know-how available on the listed techniques and tools. The block contains information on the personnel as the available *roles*, the desired *aptitude skills* for each specific role, and the required *competences*. Roles are matched directly to the techniques, while competences are matched to training. *Aptitude skills* [31] are instead soft skills as behavioral skills; which have an ancillary role in the framework but are included to present a complete characterization of personnel. More information on the roles and skills are in Section 1.4.

1.3.2 The Methodology to Exercise the Framework

The overall methodology resulting from the execution of the framework is hereby presented. The steps are the same for gap analysis of standards already in use and for the introduction of a new standard. For simplicity of the discussion, we refer here only to the last case. We assume that the standards $S_1, \ldots, S_{n-1}$ are already part of the framework, and that data on internal processes, techniques in the companies and personnel are already available. This can be done iterating the below steps for the standards $S_1 \ldots S_{n-1}$, until the dataset is up-to-date.

When a new standard S_n is introduced, the approach is the following.

Step 1. The list of standards is updated with S_n, and the corresponding traceability matrix of S_n w.r.t. internal processes is created. Table 1.1 presents a sample extract of such traceability matrix.

Step 2. The list of *techniques in standards* is updated with techniques that are mentioned in S_n; consequently, the match with *techniques in company* is updated. For example, in Table 1.2, the techniques "reviews, inspections, analysis" from the list in [27] are matched to several company techniques, as reviews, inspections, HW/SW interaction analysis (HSIA), traceability analysis. If in S_n there is a technique with no matches amongst the list of techniques in company, it is sufficient to add the same exact name to such list. As a result, a very low rating on the maturity in using such technique will be assigned in Step 4; this will be further discussed also in Section 1.4

Table 1.1 A sample extract of the traceability matrix on processes

Standard Processes (Requirements) from DO-178B	Internal Processes
SW high-level requirements comply with system requirements	Verification Process
SW high-level requirements comply with system requirements	Requirements Analysis
High-level requirements are accurate and consistent	Requirements Analysis

Table 1.2 A sample extract of the traceability matrix on techniques

Techniques in Standard	Techniques in Company
Reviews, inspections, analysis	Reviews Inspections HW/SW interaction analysis (HSIA) Traceability Static analysis

and Section 1.5. Ultimately, tools are listed and matched to the techniques in company.

Step 3. The data acquisition process gathers information on the confidence in using each technique and tool.

Step 4. Data is analyzed, and gap analysis and learning time are computed.

1.4 Dataset Structure and Population

1.4.1 Dataset Structure

With the support of the Enhanced Entity–Relationship (EER) diagram in Figure 1.2, we comment on the most relevant elements of the dataset that are

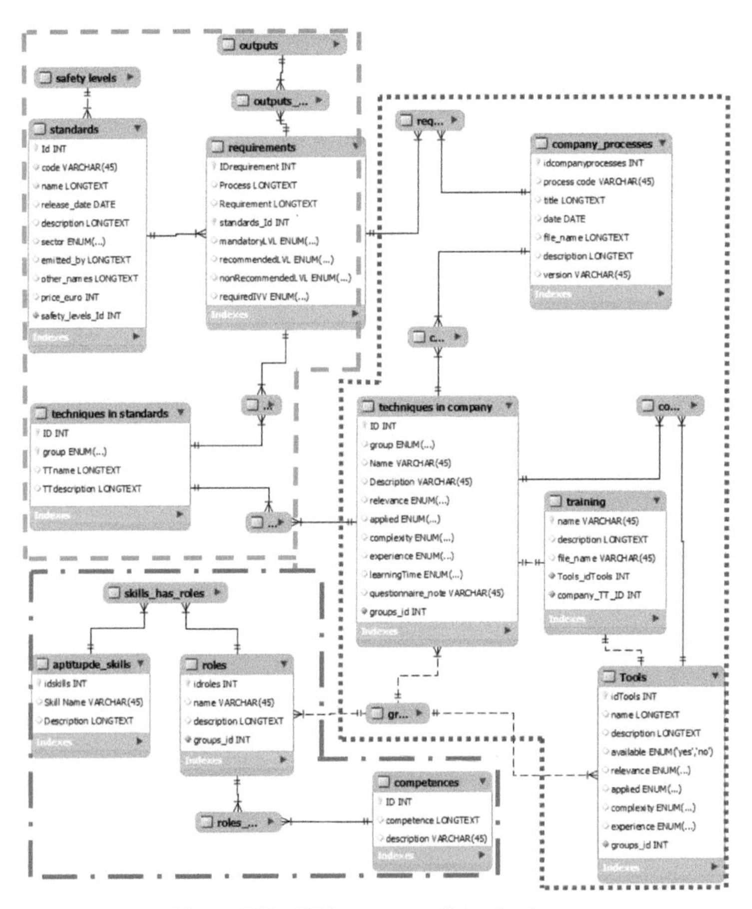

Figure 1.2 EER structure of the database.

required to exercise the framework. The diagram is organized in three areas: the first one (dashed line) contains information on the standards, the second one (dotted line) discusses internal processes and the third one (dash-dotted line) is dedicated to the definition and characterization of the personnel.

We start discussing the first area (dashed line). Table *standards* enlist the standards in use in the company including general information, for example release date, involved industrial domain, and emitting agency. Additional tables can be linked to table *standard* to annotate concepts that differs from a standard to another. As an example, the EER diagram includes the table *safety levels*, which describes the different notion of safety levels across standards. In fact, for example, safety levels are called "Software Levels" and organized in five levels in the DO-178B/C, while they are called "Automotive Software Integrity Levels" (ASILs) and organized in four levels in the ISO 26262. Other examples on safety levels can be found in [27]. Although these annotations are not deemed fundamental for the successful execution of the framework, they can simplify the execution of Step 1 and Step 2 of Section 1.3.

Table *requirements* enlist the requirements, often expressed in terms of steps of a process, described in each standard. Requirements usually suggest specific techniques: table *techniques in standards* enlist the techniques named in each standard. The *table techniques in standards* can specify if a technique is a replacement or alterative to others that are mentioned in the standard. This is useful for the mapping with the second area of Figure 1.2 (dotted line), to favor the matching of techniques in standards with those applied in a company. It is important to report recommendation level of each technique for the considered standard.

The second area includes the table *company processes*, which describes the processes available in the company. Usually, these are described in the internal documentation of a company. Table *techniques in company* enlist the techniques available. Again, such list can be extracted from the internal documentation. To perform the gap analysis, it is required to score the relevance of the technique in the daily work, its frequency of use, the complexity from the point of view of the personnel, the experience of the team in using such technique, the *learning time* (learning time indicates how much training time and hands-on-the-job time is required to gather confidence in applying a specific technique). Table *tools* contain the list of tools available in the company. For the tools table, it is required to evaluate the same attributes as above: relevance, frequency of use, complexity, team experience, learning time of the tool. Section 1.4.2 discusses how to collect such values. Finally, the table *training* enlists the training material available in the company.

The third area (dash-dotted line) is devoted to the identification of personnel. We propose the following minimum set of tables to describe the personnel, although our approach is open to improvements or adjustments in case companies offer different or enhanced characterizations of personnel.

Table *roles* enlist the different roles. Roles are related to the techniques and tools, because it is expected that people having different roles are able to apply different techniques and tools, or take responsibility over different processes. Regarding table aptitude skills, we propose from [31]: (i) behavioral skills e.g., personal integrity, interpersonal skills; (ii) underpinning knowledge i.e., knowledge on the system, required to successfully applying a technique; (iii) underpinning understanding that is general knowledge on the area of work; (iv) statutory and legislation knowledge. Table *competences*, instead, list the required competences as the number of years of experience, or the expertise in a specific topic or domain. Intuitively, table competences and aptitude skills are connected to table roles.

Relations between tables allow connecting and extracting the relevant information from the dataset. For example, the dataset can be used to verify the matching between the standards requirements and company processes. The dataset is also able to differentiate techniques that are similar but used in a different way from domain to domain; the relation of the technique to the corresponding standard is in this case fundamental.

It should be noted that terms reported in the dataset may be very general and several techniques can be matched e.g., requirements-based testing may encompass a large part of the testing activities that are performed on a system or component. The implication is that querying the dataset, different techniques applied in a company can be matched to the same technique in a standard. However this does not alter the methodology, because the different techniques available in the company are first evaluated individually, and then summarizing results are drafted, as explained in Section 1.5.

1.4.2 Population of the Dataset

We discuss hereafter how to collect the main data to populate the dataset. Some data and especially those in the first area are acquired from the documentation typically available in a company.

Regarding the second area, it is required to acquire information on relevance, frequency of use, complexity, experience, and learning time of techniques and tools. While different approaches may exist, in this chapter we propose a questionnaire that can be distributed between expert personnel to acquire anonymous data.

In this chapter, we propose the following entries and scores to rate techniques and tools applied in the company:

- *Relevance*: high relevance = 4, medium relevance = 3, limited relevance = 2;
- *Frequency of use*: often = 4, rarely = 3, and never = 2;
- *Complexity*: complex = 4, affordable = 3, and easy = 2;
- *Experience*: high experience = 4, medium experience = 3, low or no experience = 2;
    - *Learning time* (the time requested by a low-experienced worker to become able to apply a technique or tool with only periodic supervision): less than 1 month = 0.5, $\sim$1 month = 1, $\sim$2 months = 2, $\sim$3 months = 3, and more than 3 months = 4.

The possibility to select the option "*unknown*" is offered, meaning that the person was unable to decide on a rating. This option should be selected when the personnel feels that he is not able to comment on the technique or tool despite being an expert in the specific area. Also, the questionnaire is supposed to be filled only by personnel expert on safety-critical processes, so that they can adequately judge on the techniques and tools, even when they had limited opportunities to get confident with them. Ultimately, note that a questionnaire for techniques in standards is not necessary, because at least one corresponding technique in company is matched to each technique in a standard (see also Step 2 in Section 1.3).

Once all questionnaires are filled, for each technique and tool we select the following values to be computed and added in the dataset: *average, standard deviation*, *mode*, and the *number of unknowns* (number of answers in which the "unknown" option was selected). The mode can be selected instead of the average if the number of questionnaires is small or the results do not lead to a normal distribution. This may result useful in boundary cases, for example when a small subset of the personnel is very skilled on a tool, while the others do not know how to use it.

With respect to the third area (Figure 1.3) proposes a classification of the main personnel roles requested in critical software standards that can be used as reference to populate table *roles* in the dataset. Since our experience is from the aerospace, Figure 1.3 is specifically drafted having aerospace software standards in mind. The following blocks are here considered external

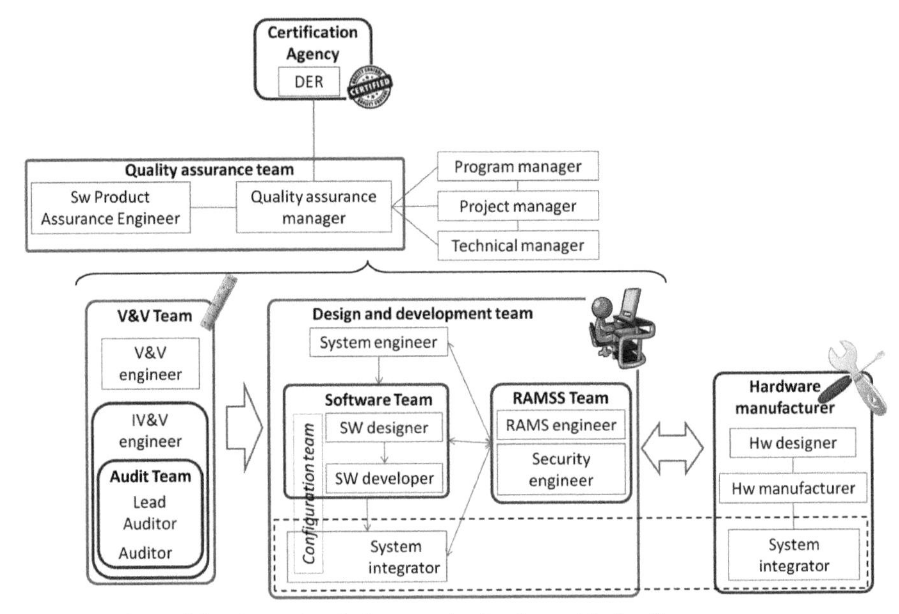

Figure 1.3 Example of roles organization in a critical software company.

to the company: Certification Agency, where the Designated Engineering Representatives, or DER, is located, Hardware Manufacturer, Independent Verification & Validation (V&V) engineer and Audit Team. This is common although it is mandatory only for the Certification Agency. System integrators are connected, because they need to interact closely for hardware–software integration.

The V&V Team and the Independent V&V Team include Test Managers and Test Engineers. The Auditor and the Lead Auditor should be included when addressing services for Independent V&V. The Design and Development Team should include also the Configuration Team, but we merge this role with Integrators, Software Designers and Software Developers. The Quality Assurance Team is in a separate group, which includes Software Product Assurance Engineers. These roles can have different aggregations on other organizations.

To verify the effectiveness of the roles subdivision, we examined the involvement of each role in the most relevant aerospace standards. We identified the personnel roles involved in the different parts of each standard, with sub-section granularity. In other words, we assigned one or more roles

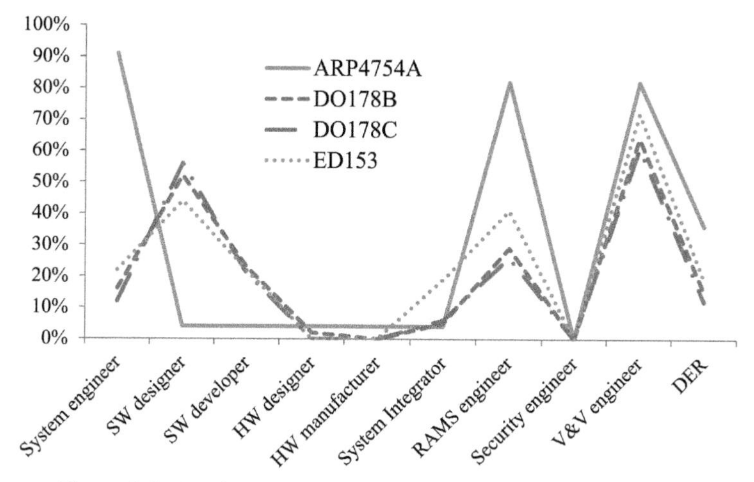

Figure 1.4 Involvement of the different roles in avionics standards.

to the requirements set contained in a subsection. We excluded introductory sections, acronyms, glossary, references sections, and Annexes.

We start our analysis from the avionics standards DO-178B, DO-178C, ED-153 [32] and ARP4754A [33]. The ED-153 applies to software that forms part of an Air Navigation System, and ARP4754A is intended for development of civil aircraft and systems, with emphasis on safety aspects. Results are depicted in Figure 1.4, showing the percentage computed approximating to the nearest integer. V&V and Independent V&V engineers are considered together due to their similar responsibilities. The managerial roles are omitted for readability as they have implicit involvement in every part of the standards. We can note that, overall, the various standards present similar percentage of involvement for the various roles. Especially, the surveyed standards have a similar percentage of V&V engineers, which ranges from 61 to 82%, and of DER. The three standards DO-178B, DO-178C and ED-153 are exclusively related to software and highly correspond for the involvement of software designers, software developers, and hardware-related roles. Instead the ARP4754A is a system-level standard and considers mainly system engineers, RAMS engineers, V&V engineers and DER, while software and hardware designers and developers have a marginal role. Security engineers are little or not considered in these standards.

We performed a similar analysis on most relevant space standards and especially software-related ones. The standards analyzed are the Galileo software standard GSWS [34], the EUROCAE Guidelines for ANS Software Safety Assurance ED-153 [32], and ECSS standards that we deemed most relevant for safety critical software design and V&V [35–38] and product assurance [39–41]. All standards showed a similar behavior, with the exceptions of GSWS considering also System integrators, and of ED-153 giving low relevance to security engineers. [36] targets mostly system engineers, and to a lesser extent RAMS Engineers and V&V Engineers. [37] and [38] instead are almost exclusively devoted to V&V Engineers. [39] and [40] are intended for RAMS Engineers and V&V Engineers.

1.5 Metrics for Gap Analysis

Once the dataset is populated, qualitative, and quantitative approaches can support the identification of the gaps and the estimation of the introduction time.

1.5.1 Qualitative Indications

We propose a qualitative analysis for the rapid identification of potential weaknesses and get an overall grasp on the results achieved. Several approaches can be identified; we propose in this chapter an intuitive one, based on a simple binary tree that can be easily built for each technique and tool.

The first four levels of the tree correspond to the attributes *relevance*, *experience*, *frequency of use*, *complexity*. The fifth level is a comment in natural language. Starting from the root, at each node, the left or right branch is selected if the score assigned to the attribute is below a threshold or not. The leaves of the tree include conclusive judgments on the technique or tool under exam.

As example, we show in Table 1.3 the binary tree that we defined for our case study. Thresholds are set as follows:

- *relevance* of the technique (for the target standard) = 3
- *experience* = 3
- *frequency of use* = 3
- *complexity* = 3

The final leaf includes a qualitative comment, resulting from the path of the tree, which may suggest the necessity of further investigation.

Obviously, this approach can be easily extended in case of additional attributes or different rating schemes that consider multiple thresholds.

Table 1.3 The binary decision diagram

Relevance	Experience	Frequency of Usage	Complexity	Qualitative Comment
≥3	≥3	≥3	any	Relevant, applied, and experienced.
≥3	≥3	<3	≥3	Relevantand large experience, but not applied.
≥3	≥3	<3	<3	Relevant, simple, large experience, but not applied. **Requires further investigation.**
≥3	<3	≥3	≥3	Relevant and complex. *Applied with little experience.* **Requires further investigation.**
≥3	<3	≥3	<3	Relevant and *applied with little experience.* **Requires further investigation.**
≥3	<3	<3	Any	*Relevant but not applied*and not experienced.**Requires further investigation.**
<3	≥3	≥3	Any	Little relevance, large experience, and applied.
<3	≥3	<3	Any	Little relevance, and not applied.
<3	<3	≥3	Any	Little relevance, but applied *with limited experience.* **Requires further investigation.**
<3	<3	<3	Any	Little relevant, and not applied.

1.5.2 Quantitative Indication

The data acquired may contain information that is not grasped during the qualitative analysis above. We define the quantities Q_1, Q_2, Q_3, Q_4 to relate relevance, (team) experience, frequency of use (called also *applied* below for simplicity), complexity, and to identify those techniques and tools that may need particular attention. Obviously several other different quantities could be identified and applied, without introducing any limitation to the methodology.

We select Q_1, Q_2, Q_3, Q_4 to seek the appropriate balance between complexity, relevance, frequency of use and team experience. The score 0 represents a balance between the different attributes; the highest it is, the highest is the necessity of further investigating the technique or tool.

Is complexity an issue? $Q_1 = complexity^2 - applied \times experience$. This quantity raises awareness of misalignment between difficulty and confidence. Q_1 is intended to heavily penalize complex techniques. A small Q_1 means that there is high confidence in the usage of a technique.

Is Experience Adequate? Q_2 = *(relevance + applied)* – *(experience* × 2). The objective of this quantity is to indicate that experience is sufficient w.r.t. the relevance and application of a technique.

Is there an overall balancing? Q_3 = *(relevance* × *complexity)* – *(applied* × *experience*). Q_3 compares the confidence in using a technique, i.e., experience and frequency of application, to the relevance and complexity of the technique. It is a summarizing quantity that relates all attributes used up to now.

Is experience justified? Q_4 = *relevance – experience.* Q_4 indicates the experience of the team w.r.t. the relevance of a technique or tool. Ideally, its target score is 0, meaning for example that a very relevant tool is applied with excellent skill; or on the opposite, that a tool recognized as almost irrelevant is also almost unknown. If Q_4 is a positive score, it indicates that a tool or technique acknowledged as relevant is not known adequately.

The case study in Section 0 reports results of these metrics for the analysis of DO-178B in a software company.

1.5.3 Driving Conclusions

The data and the results of the qualitative and quantitative analysis need to be investigated to finalize conclusions. The optimal is that for each recommended technique in the standard, one or more techniques in company are frequently applied with good experience. However, we note that often the techniques recommended in the standards can have replacement techniques, or only a subset of such techniques is actually necessary: this is further elaborated in Section 1.7.

Checks of paperwork or interviews can be a viable support to verify the gaps resulting from the above analysis. This is especially true in two cases. First, we should consider the case when different techniques can be used as alternatives to meet requirements of the standard. A gap in a technique may actually be irrelevant as far as other substitute techniques are applied with good confidence. Second, it is required to prove that the techniques are actually practiced with the skill level declared by the personnel. It is fundamental to know whether the personnel are really practicing in an effective way as declared, matching the on-paper capability of the organization with the as-practiced capability.

Whenever a gap is identified, the value of the *learning time* estimates the time required to fill the gap. The learning time indicates the effort required to train people on a technique or tool; the overall cost to cover the gap should also include the cost of tools licenses, if needed.

Finally, the *introduction time* of the standard can be estimated from all the learning times from techniques and tools where a gap is identified.

1.6 Case Study and Gap Analysis for DO-178B

The framework and methodology were applied within CRITICAL Software S.A.personnel for what concern the DO-178B standard for avionic systems. To reduce complexity, the analysis of the DO-178B we performed was limited to the sections devoted to verification and validation.

CRITICAL Software is an international information systems and software company, headquartered in Coimbra, Portugal, where our experiment took place. While CRITICAL Software works across several markets, in this work we referred to the aerospace division, which is active since 1998. In fact, it has to be noted that CRITICAL Software has relevant experience with DO-178B, applied successfully in several projects for many years. Consequently, it is evident that the objective of this case study is *not* to identify possible lacks in CRITICAL Software processes or inadequate knowledge about the required techniques, but it is to exercise the framework in a real context and verify its applicability.

Relevant data on techniques in standard were acquired from [27]. Techniques in company and tools were identified from material available at CRITICAL Software and expert involvement: this ranged from short interviews/meetings, to training material, publications, and leaflets, V&V plans for different projects, V&V reports, case studies and specific tools reports. The engineering personnel were also interviewed in order to gather the list of tools they typically use, and that may not necessarily be referred on written reports. In total, 22 Verification & Validation techniques were identified; the validation technique *testing* was further subdivided in 26 testing techniques. The number of tools identified is instead 41.

1.6.1 Matching of DO-178B Techniques and Company's Techniques

Matching between standard's and company's processes was performed by manual inspection of the standard and the company's internal processes.

For each verification and validation technique in the standard, one or more techniques were identified in CRITICAL Software processes, use cases, and V&V plans.

We summarize main results. At least one *technique in company* was assigned to each *technique in standard*. There were more than one in some cases. For example the entry "reviews, inspections, analysis" from the table technique in standard is matched to reviews, inspections, HW/SW interaction analysis (HSIA), traceability, static analysis. Similarly, the requirements-based testing amongst techniques in standard is matched to coding/unit testing, system testing, functional testing and black box testing from techniques in company.

General comments on the examples above are that i) such techniques presents significant overlaps, e.g., between functional and system testing, and ii) terms reported in the standards are often very general and several techniques can fit them e.g., requirements-based testing may encompass a large part of the testing activities that are performed on a system or component.

1.6.2 Acquire Data from Personnel

Questionnaires were filled independently by eight CRITICAL Software workers, operating as V&V, RAMS engineers or having managerial responsibilities, prevalently in the context of verification and validation and certification projects. The engineers had been selected with different experiences and expertise in order to make the questionnaires results more representative of the company level. The data were analyzed, and average, mode, standard deviation, minimum value, maximum value and number of unknowns were computed and added to the database.

1.6.3 Analyze the Data: Techniques

To favor understanding the structure of the results, an extract of the data sheets we compiled is reported in Table 1.4.

For most techniques, the standard deviation was rather limited (below 0.5) showing that despite the limited number of questionnaires, there was a good-to-high convergence of answers. Thus we preferred to use the average rather than the mode in our case study.

- *Complexity*. Less complex techniques were identified in reviews, inspections (e.g., Fagan, or walk-throughs), static analysis, traceability, code analysis, HW/SW interaction, and almost all testing techniques. Instead the most complex techniques were recognized in formal methods and modeling, with an average complexity of 3.8 (we remember from

Table 1.4 An extract of our sheet for data analysis; overall it contains 48 techniques and 41 tools. The whole data set is not reported because of its dimension and non-disclosure agreements

Technique	Relevance	Experience	Frequency of Usage	Complexity	Q_1	Q_2	Q_3	Q_4	Binary Tree	Learn. Time
Reviews	4	3.75	4	2.12	–10.48	0.50	–6.50	0.25	Relevant, applied, and experienced	Less than 1 month
Inspections	4	3.65	4	2.5	–8.25	0.75	–4.50	0.37		
Traceability analysis	4	3.5	3.87	2.25	–8.50	0.88	–4.56	0.5		
Static analysis	3.62	3.62	3.75	2.25	–8.53	0.13	–5.44	0		
				⋮						
Integration testing	3.5	3.25	3.71	3.12	–2,31	0.71	–1,13	0.25	Relevant, applied, and experienced	~1 month
Input-based testing	3.28	3.71	3.57	2.28	–8.04	–0.57	–5.76	–0.43		less than 1 month
Robustness testing	3.62	3	3.25	3.37	1.64	0.88	2.48	0.62		~1 month

Table 1.3 that the maximum is 4). Overall, the *unknowns* were very limited, with at the highest 3 for formal methods.

- *Knowledge.* Highest scores were assigned to reviews and inspections, Fault Trees, Dependence diagrams, testing. In particular regarding testing, although several kinds of testing are enlisted, a high score was assigned to all of them.
- *Relevance* and *Frequency of use.* The smallest scores for these two quantities were assigned to model checking/formal verification. In fact, these techniques have not been considered very relevant for the company business up to now. Amongst testing, security testing was considered of little relevance and seldom applied. The reason is mostly due to the standards in use, which only sparingly require security testing.

Overall, the execution of the binary tree suggested verifying 6 techniques. The one who raised the most interesting discussion is safety analysis, which resulted relevant and complex but little applied. The reason is that a proper and unified process for safety analysis does not exist, although the companies are constantly applying techniques that are part of the safety analysis. Other two techniques that are worth noting are usability testing and use case testing: they were rated relevant, and the personnel felt expert about them, but they were seldom applied. This scarce usage of usability and use case testing is not directly imputable to the will of the engineers but it is due to the characteristics of their projects. The other three techniques were identified as relevant but not applied and with limited experience; replacement techniques are typically used in such cases.

Quantitative indicators. Q_1 answers the question "*is complexity an issue?*" Q_1 score is 10.50 for formal methods and 10.87 for model checking, resulting in the highest score for Q_1. This is in line with all the above observations. Similarly, and not surprisingly, the lowest scores are assigned to reviews (–10.48) and inspections (below –10 in both cases), confirming that they were considered techniques with low complexity.

Q_2 answers the question "*is experience adequate?*" Q_2 relates experience to relevance and application of a technique. Most of the results are contained within the interval [–1.5; +1.5], i.e., near 0. This means that there is a good balance between the relevance and application of a technique, and the experience in its usage, thus not raising any particular alarm. Few techniques are slightly outside such interval. Although no techniques are significantly exceeding the interval, the worst value is registered by safety analysis; this is justified by the reason explained previously.

Instead regarding Q_3, which answer the question "*is there an overall balancing?*" we noticed that most of the techniques are in the interval [–7; +7]. For techniques outside such interval, relevant differences were identified between the couples [relevance; complexity] and [frequency of use; experience]. Most balanced scores, close to 0, are HW/SW interaction analysis (HSIA) and functional analysis (FFPA), considered in general with average scores around 3 for all attributes.

Most troublesome (high Q_3 score) is obviously when the score is a high positive value, suggesting that there is a bad feeling with a technique acknowledged as relevant and complex. Worst values are assigned to security assessment and safety analysis. Regarding safety analysis, the previous considerations hold. A different reasoning is instead applied for security assessment. Security assessment is (correctly) perceived as relevant, and this is easily motivated with the increasing attention that security is gaining nowadays. Also, security assessment is perceived as complex, because widely-accepted methodologies or techniques for security assessment of software-based systems are still failing to root in several industrial domains. Finally, standards sparingly mention security assessment, and consequently it is rarely applied in a company.

Finally, Q_4 answers the question "*is experience justified?*" It resulted that safety analysis has the highest score, meaning that although acknowledged as relevant, the personnel interviewed expressed some doubts on their team experience. This outcome is strictly connected to the previous considerations on safety analysis. Values of Q_4 significantly below 0 were not identified; meaning that overall there is a good balance between relevance and frequency of application of techniques.

Learning time. The shortest learning time was assigned i) amongst verification techniques, to reviews, inspections, traceability, static analysis, and ii) amongst validation techniques, to coding/unit testing, regression testing, input-based testing, boundary value analysis, smoke testing, ad hoc testing. Longest learning time was assigned to formal methods and model checking.

1.6.4 Analyze the Data: Tools

Tools connected to the above techniques were evaluated, although no specific issues were identified. Some tools were identified as little relevant, not applied, or largely unknown, but this was due to the fact that the tools list included also obsolete tools.

As an example, the quantity Q_4 resulted in almost all the tools as a negative value, with a few exceptions. In all cases, the value was in the interval [–1.1; 0.7]. Note that the best value, which is –1.1, was assigned to a text editor tool: it is reasonable to believe that there is a good experience in using it, although it is not fundamental because it can be easily replaced by other products.

1.6.5 Conclusive Recommendations and Feedbacks

As expected, no issues can be identified from the analysis. In general, the outcomes which suggest smaller confidence are those related to formal methods and model checking, although other replacement techniques are accepted in DO-178B and this does not really constitutes a gap in what concerns the DO-178B application.

It is worth observing that a long learning time (above 3 months) is assigned to these techniques, meaning that it is considered not easy to acquire proficiency with them. However, this is mostly due to the fact that the company has a limited focus in such activities, thus having a limited number of people skilled in the area.

The fact that Formal Methods and Modeling are not (for the particular case study) well ranked has several reasons, and specifically: (i) engineers are not prepared for these techniques from university and prior experience, (ii) they are not yet widely accepted in industry, especially from customers, (iii) they are more complex than others, and (iv) they lack appropriate tools support.

A final remark is about the techniques in standard that are grouped as number 11 in [27], that is, *similarity, service experience, failure statistics*. The corresponding techniques in company were rated poorly, mostly showing the entry "*unknown*" in the questionnaire for all attributes. A later analysis with direct confrontation with personnel concluded that the terms used for such techniques were unclear and confused the personnel involved. In fact, the questionnaire was provided to the personnel but entries not discussed in advance. The clarification allowed to verify the absence of any gap, thus solving all issues on similarity-based approaches for the verification of critical systems, with the only action of correcting techniques names in the dataset.

Finally, it is important to note that the case study was performed in a short time frame and its results might be interesting to plan ahead, estimate and have the company ready to tackle new domains and new certification challenges. It is relevant to mention that once these results have been presented to the company personnel, CRITICAL Software has taken actions to fill these gaps, and, in the frame of the European project FP7-2012-324334-CECRIS [42], processes, techniques and training material for safety analysis and for security assessments were developed. This outcome shows the direct impact that these types of analysis can have in prioritizing Research & Development within an organization. A more detailed discussion on this aspect, which we rate an important outcome of our work, is in Section 1.7.

1.7 Discussion about the Gap Analysis Framework

1.7.1 An Application to the Moving Process

This work represents a formalization of what is usually done by industries when tackling a new domain of expertise, but not always in a structure way and not always with all the required information to make sound decisions and appropriate plans. The results of this framework help to determine the actual level of knowledge and resources that can be reused instead of doing it in an *ad hoc* and less supported manner.

Discussing the specific *moving process* is not part of the paper but we cannot ignore it. Moving from one existing standard into a standard from another domain involves different factors. For example, the switch between space, avionics, railway or automotive domains involves at least cultural implications, domains specific adaptations, and a large learning process. We provide the basis to support this moving process, by identifying clear gaps, improvements and adaptations, and by providing an estimation of the effort of moving from one domain to the other based on what the company is already applying and the maturity associated to the application of those standards.

For the gap analysis or determination of where a certain company is before entering a specific new domain, it is essential to be able to properly and precisely model the new standards i.e., extracting the requirements, phases, techniques, outputs, etc. This is one of the main tasks of our work and it consists in studying the standards and modeling their contents. Then, it is also extremely important for a company to hold an internal knowledge base about their processes (e.g., in an internal quality management database), techniques (e.g., detailed plans) and tools (e.g., in the form of Software Development Plans and Verification and Validation Plans).

1.7.2 Time and Cost

Gap analysis processes are typically executed sparingly because of the required time, overall complexity, and cost. Consequently, we present an approach that can be executed with little time, effort and cost, provided that personnel with a strong background on safety-critical systems are available. As example, let us consider our case study. Once the framework and the methodology were ready, the whole case study including the population of the dataset was completed in a short time frame. Considering only time-consuming activities, the analysis of the DO-178B standard to fill table *techniques in standard* required 2 days, and the analysis of techniques and tools to fill table *techniques in company* and table *tools* required instead 4 days. It should be noted that these two tables will require only minor updates whenever the framework is exercised on a different standard. Two days were instead necessary to build relations between all tables. The questionnaires were filled in less than two hours each. Drafting conclusions, making interviews and presenting results required four more days.

Our analyses were carried out with a small number of supporting tools: the tools we used in our case study are a database, a spreadsheet tool, a text editor, and Java applications we developed in less than 600 lines of code. These Java applications allowed parsing the questionnaire, interfacing to the dataset, and building the binary tree. The artifacts produced by the framework, including those used in this work, are totally reusable for future analysis; this can be achieved simply maintaining the dataset.

1.7.3 Effectiveness and Reactions

Benefits of recovering a gap are usually acquired only after the introduction of the new standard is completed and the new market penetrated, or when novel services are sold thanks to the new skills acquired. This process is typically long and consequently the return on investment for covering a gap or applying a systematic quality assurance process is typically considered on the medium-long term [21].

However, since the benefits are evident, in all cases *the identification of gaps should be the trigger of recovery plans.* As an example, let us consider our case study. Although overall no problems were identified in the application of DO-178B, our analysis led the Research & Development of CRITICAL Software to focus on the topics of Safety Analysis procedures and Security Assessment techniques. In particular, two main actions were taken, partially supported by the project FP7-2012-324334-CECRIS [42].

First, research on approaches for safety analyses was started. We cite two published works that underline this research direction [43] discuss a measurable approach to fulfill the standard requirements but with an acceptable level of effort and within a reasonable timeframe [44] focus on techniques selection for safety analysis, aiming to provide to industries a ranked list of techniques that avoid specific types of issues.

Second, research on the interplay between safety and security was carried on, studying how security issues may impact safety [45] and walking towards the identification of threat assessment methodologies [46, 47]. A new security assessment process named STECA "Security Threats, Effects and Criticality Analysis" is currently under research, with the objective of making the company more competitive and more prepared to provide related services to the industry.

1.7.4 Replacement Techniques

It is important to consider that the standards, mostly based on a waterfall traditional V model, have requirements usually divided by lifecycle phases. For each of these phases there are proposed or recommended techniques, actions, and analysis, but not all of the listed techniques need to be applied in order to fulfill the standards requirements. We are aware that this situation has an impact on our framework, especially when determining gaps between standards: some gaps might be mitigated by other replacement techniques.

As future work, the most relevant foreseen improvement is to introduce the concept of *minimum set of recommended techniques.* Generally, the standards provide recommendations to several techniques but only a few are really required: several are proposed as alternatives. A company needs to be knowledgeable only with a set of such techniques. For example, formal methods are barely used, but all systems can get certified even without formal methods, and a similar reasoning can be carried out for fault injection. In this framework we are addressing this problem only in Step 4, once that all techniques have been evaluated individually. The future improvements of the framework will include solutions to automatically deal with this problem, introducing groups of techniques in the dataset, and adapting the questionnaires to rate the groups and not only the individual technique.

1.7.5 Different Approaches to Compliance

It should be remarked that compliance with standards may take place in two different ways: by sticking to what is recommended or by following tailoring rules. The framework usage applied in our case study directly fits the first approach. The second approach is also followed in practice, especially in the case that standards requirements are unclear and open to interpretations. In several cases, certification authorities' engineers or auditors supports this approach, helping the companies to adapt and accomplish the certification evidences.

In these cases, our framework can be successfully applied only after the tailoring rules are translated into requirements and are added to the dataset. Once this operation is completed, the framework can be exercised as usual.

1.7.6 Questionnaire Assessment and Bias

We want to depict the status of the company and the feeling of workers towards specific techniques and standards in general, considering that also most of the time the workers themselves are in charge of training personnel and transfer knowledge. The personnel are expected to have a general, broad knowledge of the techniques that are executed and on their usual relevance. In other words, as far as personnel skilled in certification of safety-critical software is available, our framework will be able to rate techniques event if personnel is not familiar with them.

However, a relevant concern is the risk of a bias in the outputs due to the personnel perception of their expertise and experience. It is intuitive to expect that engineers will report higher scores for the techniques they have experience with and actually use, and that the analysis may underemphasize important techniques that the company is unfamiliar with. This reflects a simplification from an engineering perspective, as we tend to apply only one technique or a simpler tool if it is accepted for the certification or for completing the job.

These considerations require that, when interviewing the personnel, a good assessor, or an expert engineer that deeply studied the considered standard, is present. Otherwise, the process and self-image of competence, for example personnel feeling they are much more skilled than they actually are, may introduce significant bias in the results.

1.8 Conclusions

This chapter proposed an *easy-to-use* framework and a supporting methodology to perform a *rapid* gap analysis on the usage of *standards for safety-critical software*. The methodology can be applied to new standards to be introduced or to standards that are already applied in the company as long as skilled personnel are available. The ultimate objective is to discover with reduced effort and minimal supporting tools how far a company is from having a sufficient level of knowledge to apply a specific standard. Also, the framework allows estimating the time required to cover the gaps. Our case study was executed in a short time frame, proving evidence of the intuitiveness of our solution. Results have been presented to a larger audience at the company CRITICAL Software SA, where the audience agreed that they reflect the global feeling about strengths and weaknesses, and recovery actions were taken by the Research & Development team.

References

[1] RTCA. (1992). *Software Considerations in Airborne Systems and Equipment Certification* (DO-178B/EUROCAE ED-12B).

[2] RTCA. (2011). *Software Considerations in Airborne Systems and Equipment Certification* (DO-178C/EUROCAE ED-12C).

[3] RTCA. (2000). *Design Assurance Guidance for Airborne Electronic Hardware*. (DO-254/EUROCAE ED-80).

[4] CENELEC. (2006). *Railway applications: The specification and demonstration of Reliability, Availability, Maintainability and Safety (RAMS) Part 1: Basic requirements and generic process* (EN 50126-1/EC: 2006-05).

[5] CENELEC. (2002). *Railway applications: Communications, signalling and processing systems – Software for railway control and protection systems*, EN 50128.

[6] CENELEC. (2004). *Railway applications: Communication, signalling and processing systems – Safety related electronic systems for signalling*, EN 50129.

[7] European Cooperation on Space Standardization (ECSS). (2014). Available at: http://www.ecss.nl/ (last accessed 11 November 2014).

[8] Penny, J., et al. (2001). *The practicalities of goal-based safety regulation." Aspects of Safety Management*. London: Springer, 35–48.

[9] Ceccarelli, A., and Silva, N. (2015). "Analysis of companies gaps in the application of standards for safety-critical software," in *RESA4CI workshop, Computer Safety, Reliability, and Security*. London: Springer International Publishing, 303–313.

[10] Karbhari, V. M., et al. (2003). Durability gap analysis for fiber-reinforced polymer composites in civil infrastructure. *J Compos. Construct*. 7.3, 238–247.

[11] Powell, G. V. N., Barborak, J., and Rodriguez, M. S. (2000). Assessing representativeness of protected natural areas in Costa Rica for conserving biodiversity: a preliminary gap analysis. *Biol. Conserv.* 93.1, 35–41.

[12] Brown, S. W., and Swartz, T. A. (1989). A gap analysis of professional service quality. *J. Market.* 53, 92–98.

[13] CMMI Product Team. (2010). "CMMI for Development". Technical Report, Software Engineering Institute, CMU, Pennsylvania.

[14] ISO/IEC 15504. (2004). *Information technology – Process assessment* 2004.

[15] Hanrahan, R. P. (1995). "The IDEF process modeling methodology," in *Software Technology Support Center*, New York, NY: IEEE 1995.

[16] Hanafizadeh, M. R., Saghaei, A., and Hanafizadeh, P. (2009). An index for cross-country analysis of ICT infrastructure and access. *Telecommun. Policy* 33, 385–405.

[17] El-Gabaly, M., and Majidi, M. (2003). *ICT Penetration and skills gap analysis*. Egypt: US AID's Mission in Egypt.

[18] Chinn, M. D., and Fairlie, R. W. (2010). ICT use in the developing world: an analysis of differences in computer and internet penetration. *Rev. Int. Econ.* 18.1, 153–167.

[19] Verband der Automobilindustrie (VDA). *Automotive SPICE – Process Assessment Model*, 1st edn, 2008.

[20] Margarido, I. L., Faria, J. P., Vidal, R. M., and Vieira, M. (2012). "Towards a framework to evaluate and improve the quality of implementation of CMMI® practices". Product-Focused Software Process Improvement. Berlin: Springer, 361–365.

[21] Pino, F. J., Pardo, C., García, F., and Piattini, M. (2010). Assessment methodology for software process improvement in small organizations. *Inf. Softw. Technol.* 52, 1044–1061.

[22] Mark, S., et al. (2007). An exploratory study of why organizations do not adopt CMMI. J. Syst. Softw. 80.6, 883–895.

[23] Kemerer, C. F. (1987). An empirical validation of software cost estimation models. Commun. *ACM* 30.5, 416–429.

[24] Valerdi, R., Boehm, B., Reifer, D. (2003). "Cosysmo: a constructive systems engineering cost model coming age," in *Proceedings of the 13th Annual International INCOSE Symposium* (pp. 70–82). New York, NY: IEEE.

[25] Ceccarelli, A., Vinerbi, L., Falai, L., and Bondavalli, A. (2011). "RACME: A Framework to Support V&V and Certification," in *5th*

Latin-American Symposium on Dependable Computing (LADC), 116, 125, 25–29.

[26] Rezabal, M. I., Elorza, L. E., Letona, X. E. (2013). “Reuse in Safety Critical Systems: Educational Use Case,” in *39th EUROMICRO Conference on Software Engineering and Advanced Applications (SEAA)*, 402, 407.

[27] Ceccarelli, A., and Silva, N. (2013). “Qualitative comparison of aerospace standards: An objective approach,” in *2013 IEEE International Symposium on Software Reliability Engineering Workshops (ISSREW)*, 331, 336. New York, NY: IEEE.

[28] ISO 9001:2008 Quality Management Systems.

[29] Deeptimahanti, D. K., and Sanyal, R. (2011). “Semi-automatic generation of UML models from natural language requirements,” in *Proceedings of the 4th India Software Engineering Conference*. New York, NY: ACM.

[30] Kof, L. (2009). “Translation of textual specifications to automata by means of discourse context modelling,” in *Requirements Engineering: Foundation for Software Quality*. Berlin: Springer, pp. 197–211.

[31] IET. (2007). *Competence Criteria for Safety-related system practitioners*.

[32] EUROCAE. (2009). *EUROCAE ED-153 – Guidelines for ANS Software Safety Assurance*.

[33] SAE. (2010). *ARP4754A/EUROCAE ED-79 – Guidelines for development of civil aircraft and systems-Revision* A.

[34] Galileo industries. (2004). *GAL-SPE-GLI-SYST-A/0092 – Galileo Software Standard (GSWS)*.

[35] ECSS. (2009). *ECSS-E-ST-40C – Space engineering – Software*.

[36] ECSS. (2009). *ECSS-E-ST-10C – Space engineering – System engineering general requirements*.

[37] ECSS. (2009). *ECSS-E-ST-10-02C: Space engineering – Verification*.

[38] ECSS. (2012). *ECSS-E-ST-10-03C: Space engineering – Testing*.

[39] ECSS. (2009). *ECSS-Q-ST-30C: Space product assurance – Dependability*.

[40] ECSS. (2009). ECSS-Q-ST-40C: Space product assurance – Safety.

[41] ECSS. (2009). ECSS-Q-ST-80C: Space product assurance-Sw product assurance.

[42] CECRIS. (2016). FP7-2012-324334-CECRIS: CErtification of Critical Systems. Available at: http://www.cecris-project.eu/

[43] Silva, N., and Vieira, M. (2013). "Certification of embedded systems: Quantitative analysis and irrefutable evidences," in *2013 IEEE International Symposium on Software Reliability Engineering Workshops (ISSREW)*. New York, NY: IEEE.

[44] Silva, N. and Vieira, M. (2014). Towards making safety-critical systems safer: learning from mistakes," in 2014 IEEE International Symposium on Software Reliability Engineering Workshops (ISSREW), vol., no., 162–167.

[45] Nostro, N., Bondavalli, A., Silva, N. (2014). *Adding Security Concerns to Safety Critical Certification*. ISSRE Workshops, 521–526. New York, NY: IEEE.

[46] Nostro, N., Ceccarelli, A., Bondavalli, A., and Brancati, F. (2014). Insider threat assessment: a model-based methodology. *SIGOPS Oper. Syst. Rev.* 48, 3–12.

[47] Nostro, N., Ceccarelli, A., Bondavalli, A., and Brancati, F. (2013). "A methodology and supporting techniques for the quantitative assessment of insider threats," in *Proceedings of the 2nd International Workshop on Dependability Issues in Cloud Computing – DISCCO '13*, 1–6, Braga (Portugal).

2

Experiencing Model-Driven Engineering for Railway Interlocking Systems

Fabio Scippacercola[1,2], András Zentai[3] and Stefano Russo[1,2]

[1]DIETI, Università degli Studi di Napoli Federico II, Via Claudio 21, 80125 Napoli, Italy
[2]CINI-Consorzio Interuniversitario Nazionale per l'Informatica, Italy
[3]Prolan Process Control Co., Szentendrei út 1-3, H-2011 Budakalász, Hungary

2.1 Introduction

For a company to be competitive in the market, following technologies and being updated with new trends and practices, is essential. In safety-critical domains, the introduction of new practices and methodologies is slower than in other engineering fields, since safety standards and long established practices tend to defer the adoption of new emerging technologies, until assessments and time reveal them mature and safe. Slow introduction of new methods is especially characterizing the railway domain, where the lifespan of products could easily reach decades or even a century. Now it is long time that Model-Driven Engineering (MDE) techniques and tools have been proposed, but their maturity – especially for safety-critical systems – is still debated.

Some recent surveys investigated the adoption of MDE methodologies and technologies in practice [1, 2]. They revealed the increasing adoption of MDE in industry. The technology is attractive for the development of critical systems, since it can speed up the activities of Verification and Validation (V&V), and it enables the early verification of systems, through techniques such as model reviews, guideline checkers, Rapid Control Prototyping (RCP) and Model- and Software-in-the-Loop Tests. These techniques shift the cost of development from the phases of V&V to the ones of requirement analysis and design, thus leading to benefits in terms of residual errors. Companies not performing model-in-the-loop testing find almost 30% more errors during module test [3].

Prolan Co. is a Hungarian company, which develops certified products for safety critical process control and rail signaling systems. Prolan joined the European project "CErtification of CRItical Systems" (CECRIS [4]); in its framework, Prolan started an industrial-academic partnership for the transfer of knowledge of MDE techniques from the academy to the company, with the goal of assessing their level of maturity for industrial adoption.

During this activity, it emerged the lack of well-defined processes for the development of a CENELEC SIL-4 safety critical signaling system that was suited for the real industrial needs.

2.2 Background: MDE

As for most engineering branches, advances in software engineering have always resulted from increases in the level of abstraction. Let us consider, for instance, one of the most peculiar activities of this discipline, namely computer programming: the first abstraction, i.e., the second generation programming languages – or assembly languages – were born soon after programmers had struggled with binary machine code; then came the third generation programming languages (procedural and object-oriented), that freed the programmers from low-level details of the machine, and then fourth generation languages, which added more facilities and masked recurrent problems, such as the representation of data and the interworking between heterogeneous systems. The same holds in other areas, such as operating systems, middleware technologies, and network protocols. In this perspective, MDE aims at raising the level of abstraction in software design and verification [5], and promises to change the traditional methodologies of software development.

Model-driven approaches focus on a model, i.e., on a set of specifications or representations of a system that neglect aspects that are not of interest at the current stage in a software process; the process advances transforming the model in documents, intermediate artifacts, or in the final product. The result is that MDE shifts the traditional development paradigm, based on different kinds of artifacts composed by domain experts in multiple formats, to a common formalism – the model – by which the artifacts are obtained through computer-assisted transformations. This model-centric paradigm provides several benefits, leading to increased productivity and quality of artifacts, shorter development time, and enhanced automation, which includes automatic code generation and automatic support to the software engineering activities.

Since models have always been applied at different extents in engineering problems and activities, there are many acronyms with fuzzy borders in the universe of software engineering. We refer to the terminology of Brambilla et al. [6].

When processes exploit models as support for their goals they are part of Model-Based Engineering (MBE), and we call the activities document-centric, since models are only a means to achieve the targets, but there is no particular emphasis on them. Therefore, MBE is the broadest term, encompassing all the methodologies and activities that employ models.

Model-Driven Engineering focuses on the processes where models are key artifacts of the activities (model-centric). When we restrict to considering MDE for supporting the development of systems, we can use the more specific term of Model-Driven Development (MDD). One approach of MDD is the Model-Driven Architecture (MDA), proposed by the Object Management Group (OMG) [7]. The Model-Driven Testing (MDT) is a theory of software testing that introduces concepts enabling to transform models in test-cases in order to support V&V activities. Even though MDT is not an OMG standard, it uses an OMG's standard profile, the UML-Testing Profile (UTP) [8, 9].

Model-Driven Engineering is founded on concepts of *models* and *transformations*: instead of producing (textual) documents as artifacts – requirements, design, code, test artifacts – engineers focus on models as primary artifacts.

Models are defined in (semi-)formal languages, which are typically machine-understandable and drawn with the support of tools. Other artifacts are derived through defined transformations, be they: Model-to-Model transformations (M2M) or Model-to-Text transformations (M2T) from models to textual documents, source code or testing artifacts (such as test cases and test scripts).

As argued by Kent [10], MDE can identify different levels of decomposition and can employ ad hoc or domain-specific languages for models and transformations, whereas MDA is bound to OMG's standards.

The OMG is an international open membership not-for-profit consortium grouping many IT companies and organizations around the globe. OMG first conceived MDA as a technology to overcome the interoperability problems of applications partially addressed by the CORBA standard [11]. Indeed, even if CORBA provided a good solution for the interoperability of applications, it became soon clear how it is difficult for large enterprises to standardize on different middleware platforms: enterprises have applications on different middleware, that have to be integrated even though this process turned out to be expensive and time-consuming. Furthermore, middleware systems continue to evolve and even CORBA could not be a guarantee for next decades. Therefore, MDA was proposed as a better way to reach portability, interoperability and reusability through architectural separation of concerns in the OMG vision that postulates how the myth of a standalone application or standard for developing software as well as for data interchange died.

The recent version 2.0 of the Guide to the standard [7] defines MDA as an approach for deriving value from models and architecture in support of the full life cycle of physical, organizational and IT systems. MDA became an approach to deal with complexity and interdependences in large systems, namely to derive value from modeling by defining the structure, semantics, and notations of models using industry standards.

In order to enable (automatic) transformations of models, mechanisms were introduced to reason on the models themselves: this has been done through the concept of meta-modeling, namely introducing models for modeling languages. These concepts are commons to MDE, but MDA standardized the formalisms to use, so as to have four layers of abstractions:

- M0 is the user data layer, it is the layer at lowest abstraction and the elements are concrete objects of the problem domain.
- M1 is the layer of modeling concepts. Here are the UML models of entities that abstract the user data layer, like UML classes or association. At this level are models defined by software engineers to define the requirements or architecture of the system.
- M2 is UML Metamodel, i.e., M2 defines, through UML, the syntax of UML models in M1, as well as their semantic. For instance, M2 will constraint you to do not use UML links for connecting classes but UML objects. M1 models can be seen as instances of concepts of M2 layer and, by M2, you can check consistence of your UML models.
- M3 is most abstract layer defined by OMG. At this level is Meta-Object Facility (MOF) language. By MOF OMG can define syntax and semantic for meta-languages. In the MDA, MOF enables to define transformation rules among different models (of M1 layers) that are compliant to different meta-models (of M2 layers).

Using its modeling infrastructure, it is possible to define rules to transform models into other models (M2M) or model into text (M2T). With M2T transformation, MDE refers specifically to that kind of transformation that produces source code (or other textual documents) from models.

2.2.1 MDA Viewpoints and Views

Model-Driven Architecture starts with the well-known and long-established idea of separating the specification of the operation of a system from the details on how that system uses the capabilities of its platform. MDA enables to specify a system independently from the platform that supports it, and to transform the system specification into one for a particular platform.

A *viewpoint* specifies a reusable set of criteria for the construction, selection, and presentation of a portion of the information about a system, addressing stakeholder concerns [7]; in other words, a viewpoint defines the abstractions to adopt to focus on particular concerns within the system. A *view* is a representation of a system that conforms to a viewpoint [7].

In MDA terms, abstraction eliminates certain elements from the defined scope and may result in introducing a higher-level viewpoint at the expense of removing detail. A more abstract model encompasses a broader set of systems, whereas a less abstract model is more specific to a single system or restricted set of systems. One important capability of MDA is the automation that provides for the transformation between levels of abstraction by the use of patterns.

Model-Driven Architecture specifies three viewpoints, which offer levels of separation of concerns to realize a system. The three viewpoints are:

Computation Independent Viewpoint (CIV). The computation independent viewpoint focuses on the environment of the system, and the requirements for the system; the details of the structure and processing of the system are hidden or as yet undetermined;

Platform Independent Viewpoint (PIV). The platform independent viewpoint focuses on the operation of a system while hiding the details necessary for a particular platform;

Platform Specific Viewpoint (PSV). The platform specific viewpoint combines the platform independent viewpoint with an additional focus on the detail of the use of a specific platform by a system.

The recent version of MDA standard [7] reduces the emphasis on the CIV, and defines a platform as a set of resources on which a system is realized. This set of resources is used to implement or support the system. For instance, a platform can be the organizational structure or a set of buildings and machines (in case of business or domain platform types); or operating systems, programming libraries, and CPUs (when considering computer hardware and software platform types).

A platform model also specifies requirements on the connection and use of parts of the platform, and the connections of an application to the platform. Example: OMG has specified a model of a portion of the CORBA platform in the UML profile for CORBA. This profile provides a language to use when specifying CORBA systems. The stereotypes of the profile can function as a set of markings. A generic platform model can amount to a specification of a particular architectural style.

Considering the previous views, MDA defines the Computation independent Model (CIM), the Platform Independent Model (PIM), and the Platform Specific Model (PSM). MDA refines CIM in PIM and in PSM using model transformations during development process.

2.3 The Maturity of MDE

Several surveys analyzed the diffusion and the benefits of Model-Based and Model-Driven techniques and technologies into industrial practices, after 30 years from the introduction of the first MD tools on the market. However, these analyses are still not enough to get a complete picture about the state of the MD practices. Indeed, an aspect that is often neglected by these surveys is that the utilization of MD techniques is tightly dependent on the domain, which influences the demands of the users as well as the stability of the environment and the availability and maturity of the supporting tools. The domain of embedded systems, for instance, has seen the diffusion of sophisticated MD tools such as Matlab Simulink or SCADE, that keep evolving in the offered functionalities since they were introduced on the market, many years ago. However, if we consider all other domains, we can see that the adoption of MDE in software companies differs from that in the domain of embedded systems: although MDE is always perceived beneficial, the benefits are not as evident as in the embedded system industry.

In general MDE seems not completely mature yet, and the feasibility of its adoption partially debated, with tools not enough stable integrated, and much of the MDE potential yet to be demonstrated. Summarizing the various observations in the surveys of past years in industry, we may conclude that:

- *MDE is spreading in industry, but it is still far to be pervasive.* It followed the concurrent evolution of modeling languages (such as UML) and of related tools: in 2005 practitioners were using MBE for conceptual modeling [12], in 2008 model-centric approaches were perceived better than code-centric ones in most of tasks [13], in 2010 and 2011 MDE has been observed in a wide range of application domains [14–21], despite there are many problems and no general and common consensus on these approaches;

- *Models are mainly used for design and documentation*, while the benefits of advanced techniques (such as code generation, test case generation, or model animation) are lowly exploited: models are introduced mostly as an enabling technology inside the process, to enable business that otherwise would not be possible [14–16];
- *UML is gaining popularity, but support tools are not enough mature yet to build toolchains meeting the specific needs of companies*: they are considered one of the biggest problem by the industry, that is worried about ease of their usage, the vendor lock-in problem, and the interoperability among different tools [18–21];

Model-Driven Engineering depends on the business domain and on organizational factors, and its adoption requires changes in the personnel skills, the software processes, and the company practices. MDE demands for special skills and for changes in the roles of developers and software engineers: retraining programmers to think at a higher level of abstraction can reveal a difficult task. These aspects have not been well addressed so far, and the current approaches do not adequate to the people, but the people have to adapt to them.

A partially different scenario is observed in the domain of embedded systems, where we can draw the following picture:

- Model-based techniques are widely adopted (almost pervasive in automotive domain), and models are used not only for informative and documentation purposes but they were the key artifacts of the development processes [1, 2].
- The needs for introducing models was mainly for shorter development time, and to improve reusability and quality, whereas less than half had the need to introduce models to exploit formal methods, or because they were required by the standards [1, 2].
- The activities of V&V had a huge impact by their adoption in the automotive domain [3]: the automotive industry was used to exploit model-driven approaches for the early verification of the systems, by techniques such as model reviews, guideline checkers, RCP and

Model- and Software-in-the-Loop Tests, that lead to better quality, reduced development time, due to the shifting of the costs to the phases of requirement analysis and design;

- According to [22], UML is not used widely due to short lead-time for the software development, or lack of understanding or knowledge of UML models; however this survey, limited to MDE/MDA in the Brazilian industry, does not agree with [1, 2] targeting the European industries of embedded systems. These authors found that the majority of survey participants were using Matlab/Simulink/Stateflow, followed by Eclipse-based tools. The most used modeling languages were the OMGs ones (UML and SysML);
- As for generic software companies, in the top shortcomings identified there are the scarce interoperability and usability of tools, and the high (initial) effort to train developers [1, 2].

Why was the diffusion of MB and MD techniques different in the embedded systems domain with respect to other application areas? We claim that this is due to:

i. The different weight of the activities in the development process (more emphasis on design and implementation for generic software systems; more emphasis on analysis and V&V for embedded systems);
ii. The parallel evolution of the code-centric technologies that are available for the development, which raised even more the level of abstraction during the design, and simplified the way the systems are implemented. The hypothesis partially reflects the different focus on the adoption of models in the two domains, since there is more emphasis on design and documentation in the general market, and on the V&V techniques for the embedded systems.

The cited surveys identify the current state of the adoption of MD techniques in industry by collecting the opinions of the practitioners on the benefits and drawbacks of model-based and model-driven techniques. However, besides these quantitative data, there is the need of empirical studies that analyze qualitatively and critically the merits and faults of model-driven approaches. Indeed, the success or failing factors of MDE are still unclear, and more research is needed [23].

A systematic review of empirical studies on MDE from 2000 up to June 2007 was performed by Mohagheghi and Dehlen [24]. They show that MDE can effectively reduce the cost and development time, however this depends on the grade of adoption in the development process: a success story is the one of Motorola [25, 26], that used MDE for more than 15 years in a wide spectrum of activities, ranging from protocol implementations up to handheld devices or network controllers; they experienced an increase in quality and productivity (ranging from 1.2× to 8×) and an approximately 33% reduction in the effort required to develop test cases.

Motorola could achieve these results within a mature process that was supported by own-made translators and tools for the model exploitation. Indeed, one common issue of MDE is the absence of well-defined processes [24, 27, 28], as the application of MDE requires changes in the activities, corporate culture and skills of the employees: many software engineering methods are not fitted to use models as main artifacts, and the environments seems not mature enough. Some previous studies attempted to apply pre-existing processes to MDE, or to create own ones, but MDE shifts the importance of many activities to (automatic) transformation rules, and change consolidated development process is not a naive task. The study [29] reports a successful introduction of a MBE process after 4 years and three projects had been defined and consolidated: there is the need to look beyond the technical benefits of a particular approach to MDE and instead concentrate on social and organizational issues [16].

Moreover, the process becomes a more difficult problem in safety-critical domain, where compliance with certification standards poses additional requirements on the methodologies for product life cycle. For these kind of systems, the major part of costs are for the activities of V&V, so rigorous and well-assessed techniques have to be integrated within the development process for the early detection of faults and to guarantee the quality of the product. In addition, non-functional requirements, such as safety, reliability and timing requirements, are a primary concern that have to be taken into account by these processes: current MDE methodologies do not cope with stringent functional requirements and qualities in current systems, i.e., the ability of these approaches to adapt to rapidly changing hardware and implementation platforms that are highly complex [23].

Parallel to the challenge of the product life cycle, there is the open problem of the supporting tools: they are not mature yet, and influence most of the adoption of MDE. Moreover, the vendor lock-in problem is also perceived as a problem, and the companies prefer to adopt open source solutions or to develop their own tools. Indeed, the tools are not well usable, do not interoperate between themselves, do not keep in synchronization the models at different level of abstractions, are not flexible to collaborative working, and are not suited with the adoption of different models and modeling notations [23]. Thus, model-driven processes have to carefully consider the problem of defining the toolchain for supporting the activities.

2.4 A Model-Driven Methodology for Prolan

In the period when CECRIS started, Prolan was developing the next generation of railway interlocking systems, and in particular the first product of this generation, the *Prolan Block* (PB), a safety-critical system for railway interlocking that must be CENELEC EN 50126, EN 50128 and EN 50129 SIL-4 certified.

The system is deployed alongside railway segments, which are named *blocks*. Each block is equipped with a PB, with sensors for detecting incoming and outgoing trains (these sensors are the axle counters), and with semaphores that are part of the signaling system. The PB manages the block (Figure 2.1), receiving data from sensors, and properly setting the semaphores according to its internal state.

The interlocking is realized by the overall distributed system that consists of interacting PBs, which must ensure that no collision will happen on the railway, directing the train movements by proper sequences of signals. For instance, according to the specific regulations, the yellow lamps can indicate that the next block's semaphore is red because there is an obstacle (e.g., a train) in the block after the next (e.g., there is a train two semaphores ahead).

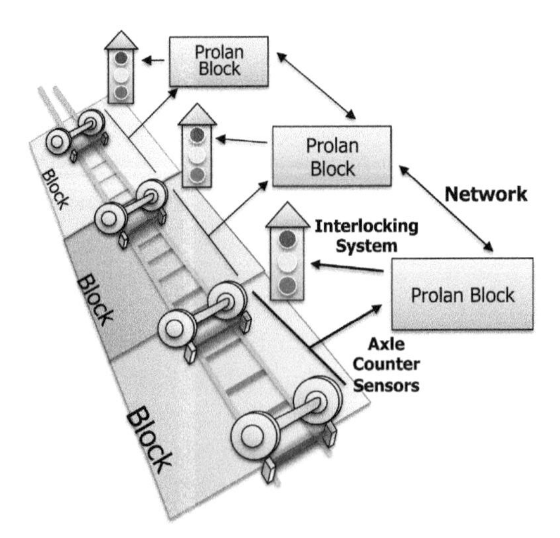

Figure 2.1 A representation of the Prolan Block and its operating environment.

Prolan was interested in understanding the potentialities that model-driven technologies could give for the development of the PB and for the other products of the same generation. Indeed, small and medium size enterprises like Prolan are interested in model-driven technologies but there are barriers to their introduction, for the deep changes that these require into the organization and in the current industrial practices. Indeed, for the adoption of MDE Prolan needs to carefully rethink and redesign its current product development life cycle, that currently complies with the railway standards CENELEC EN 50126, EN 50128 and EN 50129, as well as the skills of the employees, even if no proven-in-use model-driven lifecycle for this domain is available and supported by long-term evidence.

The traditional Prolan development life cycle follows the V-Model and is compliant with the European railway standard EN 50128. The activities of the CENELEC V-Model process can be grouped in those concerning development, that are on the left side of the 'V', and those focusing on V&V, that are on the opposite side as it can be seen in Figure 2.2. The activities of V&V require planning stages that are performed before their actual execution: these planning stages are carried out during design.

Besides the activities in the V-Model, CENELEC EN 50128 also prescribes requirements on the documents produced at each stage, as well as on the project organization. For instance, if we consider the highest integrity level (SIL-4), distinct people have to test, verify and validate the product, in order to cross-check their work. The phases adjacent to the 'V', the Software Planning and Software Assessment, aim at tuning and assessing the activities of the life cycle, defining the tasks to be performed during the process and checking that the product and all artifacts satisfy the requirements and comply with the standard.

To gain experience on model-driven technologies, Prolan started a collaboration with CINI in the framework of the CECRIS Project to develop a development process enhanced with model-driven approaches.

Since Prolan wanted a concrete and feasible option for replacing its current methodology, the researcher proposed a development process backed to Prolan's traditional process. This solution minimizes the impact of the change on the organization, and is also compatible with the safety standards pursued by the company. The adaptation of the development processes of Prolan to MDE focused on core phases of the CENELEC V-Model, starting from the System Development Phase up to the Software Validation Phase, i.e., on the Software Development Life Cycle (SDLC).

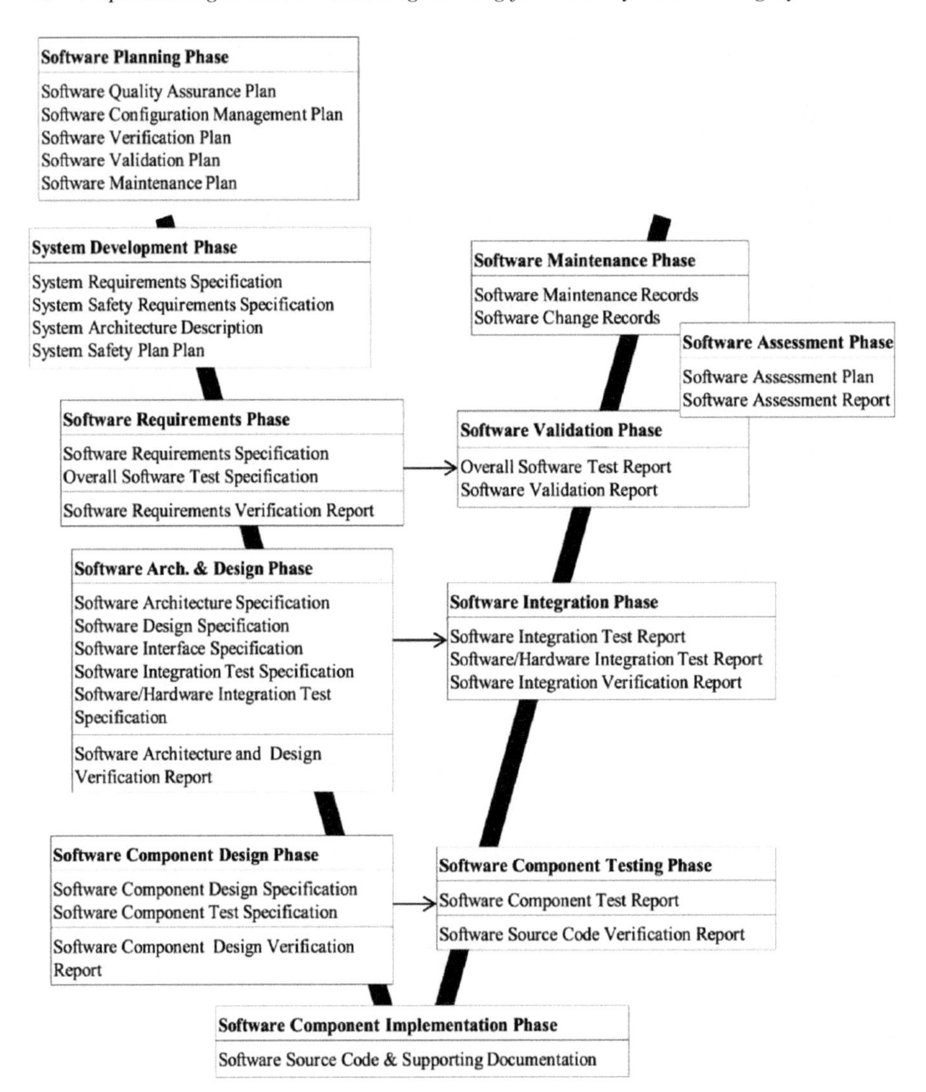

Figure 2.2 Software Development Life Cycle according to EN 50128.

The proposed model-driven V-Model lifecycle is shown in Figure 2.3: it is composed of a left, center and a right part; the title lines of the boxes refer to the SDLC activities performed by Prolan, according to CENELEC EN 50128 standard. For each activity, the boxes contain the models produced, and the formalisms used. Arrows represent dependency between the artifacts. The Component Design also depends on the Component Verification Design if it exploits the test model to early detect faults.

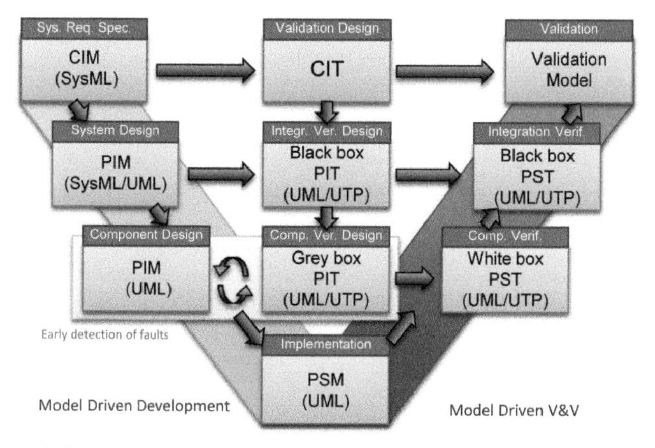

Figure 2.3 The adapted model-driven V-Model life cycle for Prolan [31].

On the left there are forward engineering activities (system analysis, design and implementation); the phases in the center are for V&V planning, while on the right side there are the activities of V&V execution. The CENELEC V-Model life cycle adopts implicitly different viewpoints on the system for each level of the 'V': the top level focuses on the system as a whole, the level below uses a viewpoint on the system architecture, then it considers the components and their internal design; finally, the lowest level of the 'V' sees source code details. These abstractions are used on both sides of the V-Model, for development and V&V.

The activities are assigned to a number of roles that comply with the CENELEC EN 50128 standard. We consider the following roles and responsibilities:

- The **Requirements Manager** is responsible for specifying the software requirements. (S)he shall be competent in requirements engineering and be experienced in application's domain (as well as in safety attributes);
- The **Designer** transforms software requirements into a solution, defining the system architecture and developing component specifications. (S)he has be competent in the application area, and in safety design principles;
- The **Implementer** transforms design solutions into data, source code or other representations to create the product software artifacts. (S)he has to be competent in engineering of the application area and implementation languages and supporting tools;
- The **Tester** develops the test specifications, and performs the test execution. (S)he has to be competent in the domain where testing is carried out;

- The **Integrator** manages the integration process using the software baselines, developing the integration test specification. (S)he has to be competent in the domain where component integration is carried out.

All these roles require advanced modeling skills, and experience with MDE, as well as with the adopted formalisms and tools.

The process starts with System Requirements Specification, by defining the system environment and software requirements. Then, System Design and Component Design are carried out. The former defines a high-level system architecture, identifying the hardware-software interface, and the components interfaces. Requirements are then allocated to components, and the Designer specifies their responsibilities and expected interactions. Finally, in Component Design the Designer completes the components with the internal design, and the Implementation concludes the development.

For enabling forward engineering to model-driven technologies, we define in three stages a CIM, a PIM, and PSM, following the MDA principles.

The V&V planning activities (Validation Design, Integration Verification Design, and Component Verification Design) have been isolated at the center of the V-Model. They are followed by the ones of V&V execution that are performed on the right side of the 'V', i.e., Validation, Integration Verification and Component Verification. For instance, Validation Design produces the Overall Software Test Specification after the System Requirement Specification. Then, the actual validation is performed in the Validation activity, at the end of the 'V', after Integration Verification, to assess the product conformance to requirements.

For the phases of V&V, we propose a model-driven methodology based on the MDA abstractions: the planning phases use Platform Independent Test Models, whereas the execution phases build Platform-Specific Test Models. In fact, the V&V execution phases on the right side of the 'V' benefit from the availability of the implementation, which constrains the technological platform.

Using this methodology, Prolan aims at improving the reuse of artifacts of design and V&V, supporting most of activities of the life cycle with model-driven approaches. Prolan wanted to evaluate the adoption of OMG standards, i.e., SysML [30] and UML, to be open to multiple tools and promote the interoperability of the models. It is worth to note that custom profiles can be introduced in the process to potentiate the automatic generation of artifacts throughout the whole SDLC, thus reducing the manual efforts.

We remark that since Prolan's products must undergo safety certification, one of the main requirements of the methodology is to exploit model-driven technologies for supporting multiple activities of V&V. Indeed, the proposed process is open to multiple forms of V&V, and includes techniques of early system validation, through the definition of the Computation Independent Test (CIT) model.

2.4.1 Experimentation within A Pilot Project

Prolan started a pilot project on a subset of requirements for the Prolan Block, in order to assess the benefits and drawbacks of the model-driven technologies.

2.4.2 System Requirements Specification

At this phase, the Requirements Manager defines the system and the specification of software requirements. We defined a CIM starting from the high-level system specification.

The CIM models requirements, and the relations between them, in SysML, because the language turns out particularly suited in this phase due to the Requirement diagram, the Use Case Diagram, and the Block Definition Diagram. In particular, SysML Requirement Diagram is useful to display textual requirements, and their relationships, and to trace them with other modeling elements.

Prolan built in the pilot project a CIM using MagicDraw [32], a modeling tool created by No Magic. Functional and non-functional requirements of the system were described using requirement diagrams meanwhile the system context was described with block definition diagrams. Modeling using the MagicDraw tool was introduced in an earlier phase of the CECRIS project in the framework of the knowledge transfer with Budapest University of Technology and Economics. Using models to capture requirements increased requirement quality significantly because the graphical representation made it possible to overview complex systems as well as the constraints of the modeling environment forced the engineers to create consistent requirements. We found it convenient to separate functional and non-functional requirements in different groups as well as to mark derived requirements using the refinement relationship.

An example of functional and non-functional requirements modeling can be seen in Figures 2.4 and 2.5. As it can be seen from Figure 2.6, the PB system is connected to a Radio Block Center (RBC) to a PB Human Machine Interface (HMI) to a Station Interlocking System and to track occupancy

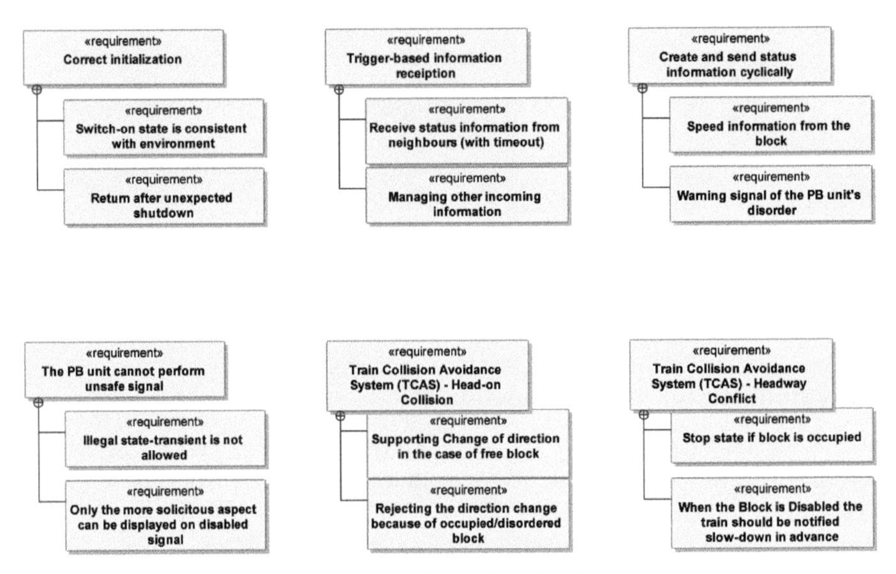

Figure 2.4 *Prolan Block* (PB) functional requirements.

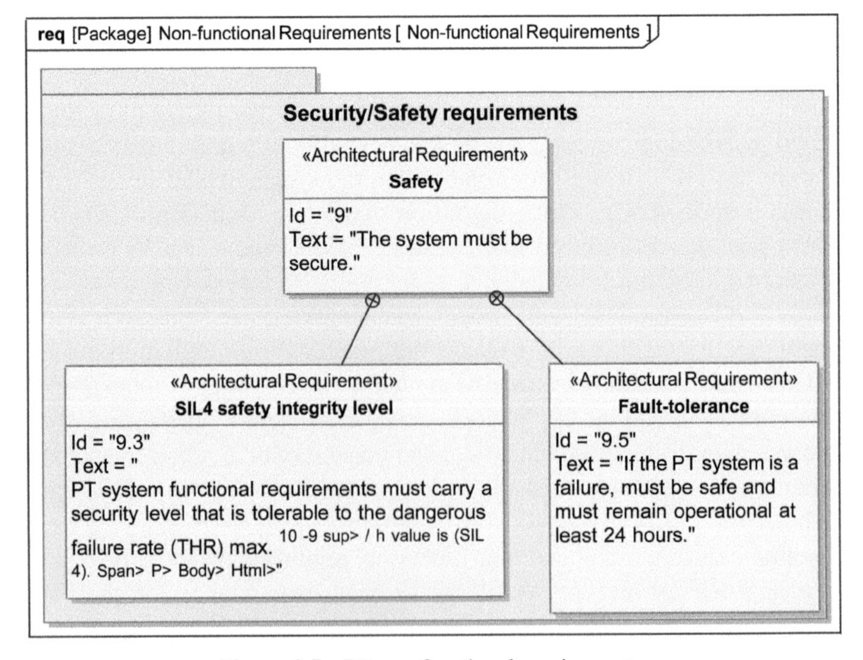

Figure 2.5 PB non-functional requirements.

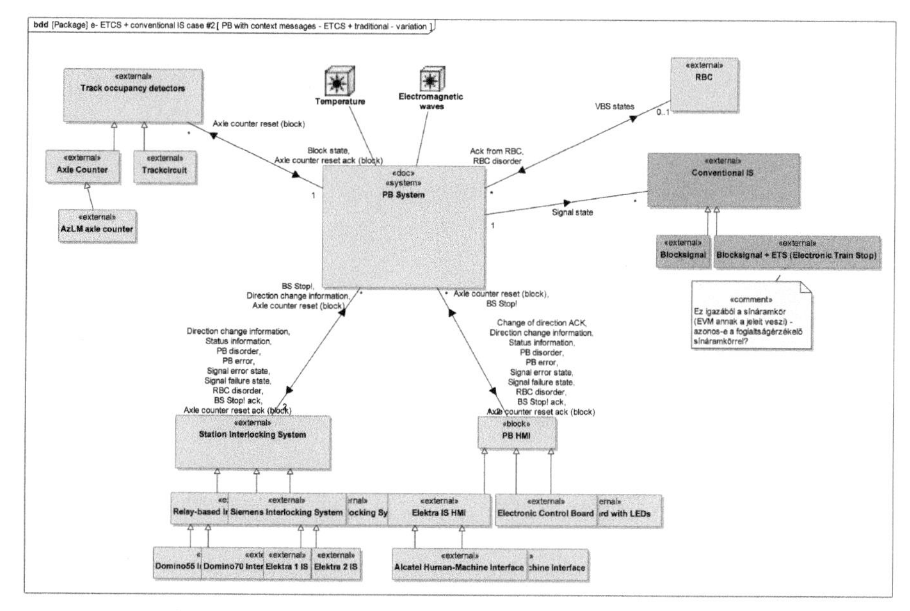

Figure 2.6 BDD diagram showing the environment of the PB.

detectors. In the BDD diagram not only the related components but also the multiplicity as well as the exchanged information and signals could be visualized.

Not only other actors, but also their relations and compositions were modeled with the BDD. Use case diagrams were created to describe in which functionalities the actors are involved (Figure 2.7).

One use case of the PB HMI is to receive the status of the PB and display it. Another use case is to reset the track occupancy detectors in case the operator activates the axle counter reset.

High-level functionalities of the system defined by functional requirements and use cases are further detailed by behavioral diagrams: state machine diagrams, activity diagrams and sequence diagrams. Requirements coming from the railway domain like the description of the semaphore's behavior (in Figure 2.8) are primarily introduced into the CIM.

Active support from the CINI side was necessary in modeling the CIM, because the technology was relatively new to the requirement engineers of the company. During this phase 6 SysML Requirement diagrams; 12 SysML Block Definition diagrams; 41 Use Cases diagrams; 6 State Machines diagrams; 29 Activity diagrams; and 33 Sequence diagrams were created.

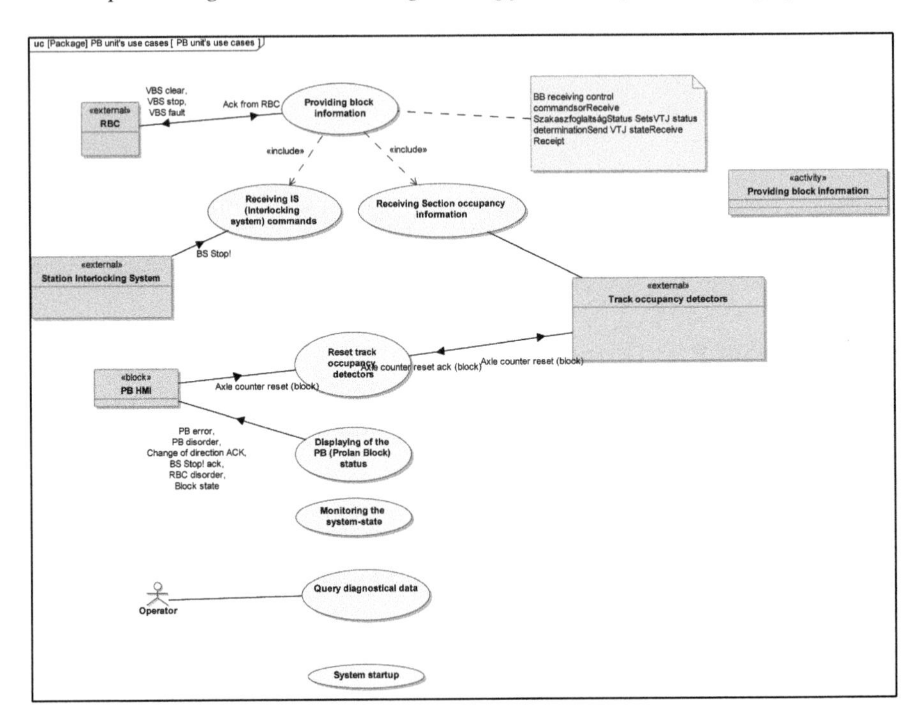

Figure 2.7 Computation Independent Model (CIM) use case diagram for the Prolan Block.

2.4.3 System Design

In System Design, the Software Designer builds the Platform Independent Model to define the software architecture, the interfaces between the components, and between the components and the overall software. To this end, (s)he uses structural diagrams, such as Component and Class Diagrams, and assigns the requirements to the system components. Since the viewpoint is platform independent, the interfaces are independent of any technological platform.

In the pilot project, Prolan used UML component and class diagrams to model the high-level architecture of the PB.

The PB design comprises five components (Figure 2.9): *Prolan Block Core Logic*, *Track Occupancy Detector*, *Network Communicator*, IS Controller, and *HMI Controller*.

By communicating with the axle counters, the *Track Occupancy Detector* notifies to the system events such as “a train entered the block” or “a train has left the block”. It also manages device failures, notifying exceptional conditions.

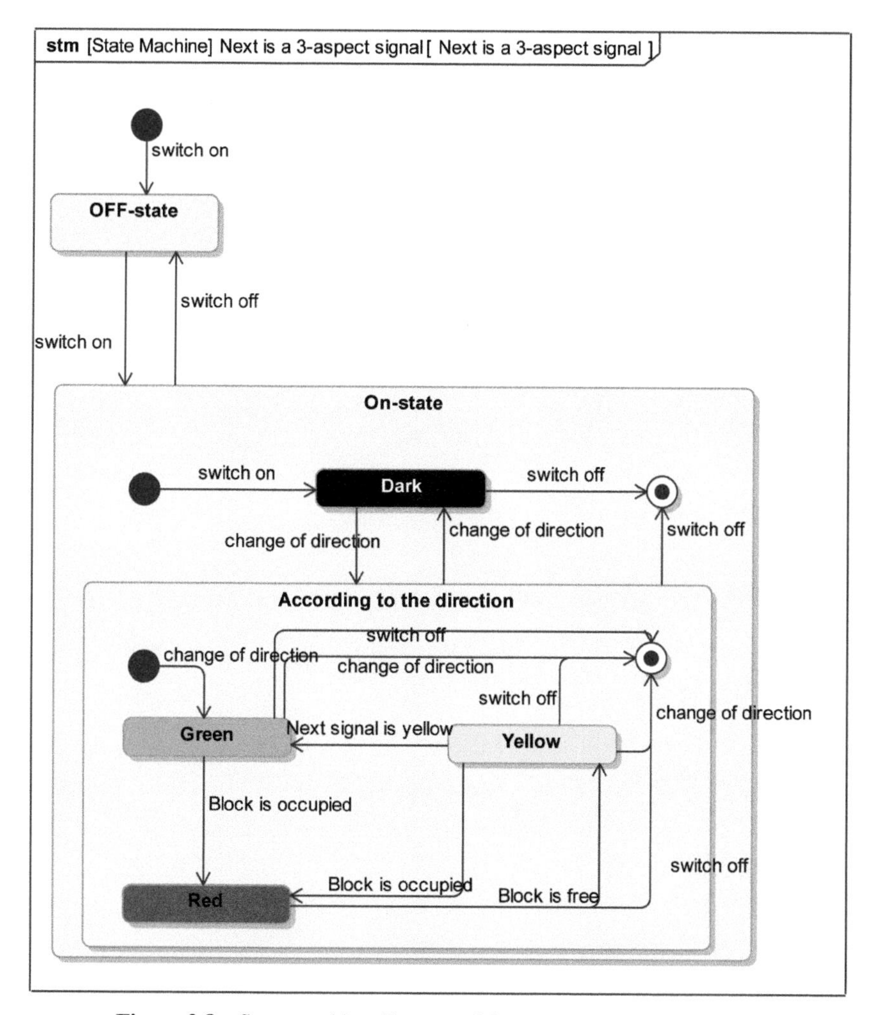

Figure 2.8 State machine diagram of the semaphore behavior [31].

The *IS Controller* interacts with the semaphore, setting the proper aspect and coping with malfunctionings. Similarly, the *NetworkCommunicator* uses the network, for interacting with adjacent PBs, and the *HMIController* manages the human-machine interface.

Finally, the *ProlanBlockCoreLogic* sets the interlocking systems according to its internal status and by collaborating with the other four components.

The components' interfaces were defined following guidelines to keep them as much as possible UML standard and platform-independent. For

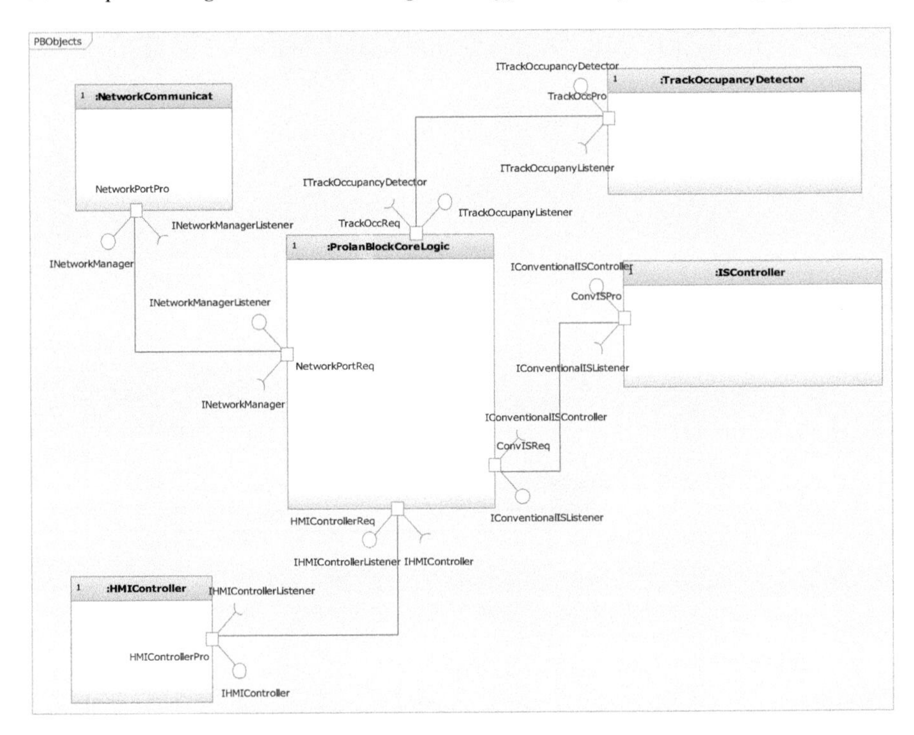

Figure 2.9 High-level system architecture [31].

instance the services of any middleware or library have been defined in terms of abstract interfaces, and the data types of the variables neglected one specific programming language.

2.4.4 Component Design

The next phase is Component Design: here the Designer refines the PIM, to specify the internal design of the components. (S)he identifies all lowest software units, fully detailing their input and output, and specifying algorithms and data structure. The PIM becomes complete, and can be runnable and object of simulation.

In the pilot project, Prolan defined the PIM using IBM Rhapsody Developer [33] (hereinafter: Rhapsody), but following the guidelines to build a model platform-independent. Indeed, the tool does not allow a clear separation between a PIM and a PSM, thus we avoided to insert C++ code, and adopted UML compliant syntax where possible.

Only few parts could not be specified in a platform-independent style. However, these parts were specified in C++, to exploit the model animation feature of Rhapsody. Indeed, Prolan was interested in this feature as a technique of early fault detection: Rhapsody's model animation generates an instrumented implementation of the model that allows to observe at runtime the program execution. This feature was useful and valuable for getting an immediate feedback on the design.

By Rhapsody Panel Diagrams, we drew a graphical user interface bound to the model that enabled to generate and receive model events at runtime. Of course, the execution can be also followed on behavioral diagrams, e.g., state machines or sequence diagrams.

2.4.4.1 Implementation

Implementation phase deals with the production of software that is analyzable, testable, verifiable and maintainable. Following MDA, the PIM is refined into one (or more) Platform Specific Models that are bound to target platforms. The PSM adds low-level implementation details. For instance, a PSM binds data and interfaces to the target OS and middleware chosen for the instantiation of the PIM.

Using IBM Rhapsody, Prolan set tagged values and other parameters to enrich the PIM with information platform-specific details, to specify how to translate the association (e.g., by static or dynamic arrays), what is the clock for the scheduler of the state machines' event queues, and other parameters. These are used by Rhapsody for the automatic translation of PSM into code.

Considering the requirements of the PB, Prolan specified that the generated code cannot use dynamic memory and that the variables have to be initialized at runtime, due to the lack of memory isolation on Prosigma, the technological platform for the PB. The platform specific code can also exploit the round-trip code feature of MDE tools:

1. By automatic code generation, packages, code skeletons, make files and other artifacts are automatically model-to-text produced;
2. The Implementer fills the code skeletons with platform-specific details, using the support of modern development environments (such as Eclipse);
3. By code round-trip, the model is automatically augmented with the information written manually by the Implementer in the source code.

According to the requirements, there are many options for the translator that include the execution framework at runtime. For instance, Rhapsody offers two C/C++ frameworks: IBM Rhapsody Object Execution Framework (OXF), and IBM Rhapsody Simple Execution Framework (SXF). The latter is dedicated to embedded systems and safety-related development: qualification kits support the certification of the automatic generated code for several standard (including ISO 26262, EN 50128 and recently DO-178B).

The translation of our PSM in C++ source code generated around 7.5 thousands of lines of code for a platform using a conventional OS, 7.3 thousands for target platform using a commercial Real Time Operating System (VxWorks), and 5.9 thousands C lines of code for an embedded systems not using an OS. We used the SXF framework in the last code generation.

2.4.5 Validation Design

For Validation Design, we propose a model, named CIT Model, to specify the behavior of the actors and of the environment. CIT can be used for designing validation tests, e.g., as UTP sequence diagrams representing the interactions of the actors with the system.

Prolan decided to does not build a CIT model for the Prolan Block, and the benefits of CIT modeling have been experimented on another system, the Prolan Monitor, that is discussed in the next section.

2.4.6 Integration Verification Design

During Integration Verification Design, the Integrator realizes integration tests to show that components behave correctly when integrated together. The expected behavior of the components is independent from their inner design, thus we refer to this model is named Black Box Platform Independent Test Model (BB-PIT).

BB-PIT provides static and dynamic views of the system's components, and supports functional testing for unit/integration/system verification (Figure 2.10). The static description supports the generation of test harness, such as stubs and drivers for unit and integration testing. The dynamic description supports the generation of test suites and test cases.

The components' behavior seems modeled twice, in PIM and BB-PIT. However, the two models have different purposes: the first specifies how to build the system, and represents the specification that an actual implementation must comply with; the second describes the expected behavior in a way to verify its correspondence between requirements and implementation (e.g., by test cases).

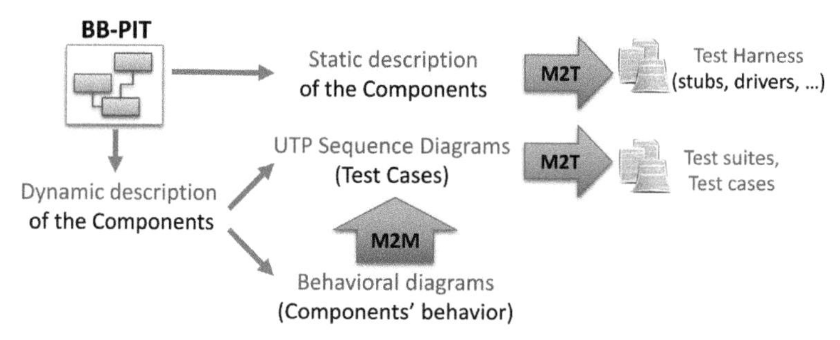

Figure 2.10 The transformations of the BB-PIT.

For tests specification, UML-UTP Sequence diagrams are less error-prone than textual notations, and it is easier to derive test cases for multiple target platforms (such as TTCN-3 and JUnit), enhancing reusability and maintainability.

In the pilot project, Prolan adopted Conformiq Designer (from here onward: Conformiq) [34] to generate automatically test cases from the BB-PIT. However, since Conformiq is not fully compliant with UML, the behavior was specified in QML, the language used by the tool. As adequacy criterion we used the requirement coverage, using the requirement traceability offered by the model-driven tools.

In total, Prolan achieved the full coverage of requirements generating 21 test cases for the *ProlanBlockCoreLogic*. Test cases were exported to JUnit from sequence diagrams (Figure 2.11), and the tool also provided us with the traceability matrix correlating test cases with the structural features they cover (states, transitions, requirements).

We also assessed the test harness generation from BB-PIT. Conformiq required to write the SUT adapter to let the testing framework interact with the system. Instead, Rhapsody offers the Test Conductor Add On [35], that automatically generates the testing harness (including the drivers and stubs), starting from model design diagrams. Within Test Conductor we could execute test cases directly in Rhapsody, observing the effects, and following the behavior of the SUT by means of sequence diagrams.

2.4.7 Component Verification Design

In Component Verification Design tests have to confirm that components perform their intended functions. Here, we define the Grey Box Platform Independent Test (GB-PIT) Model, which is used for verification by the internal view of the components. Following this flow, engineers

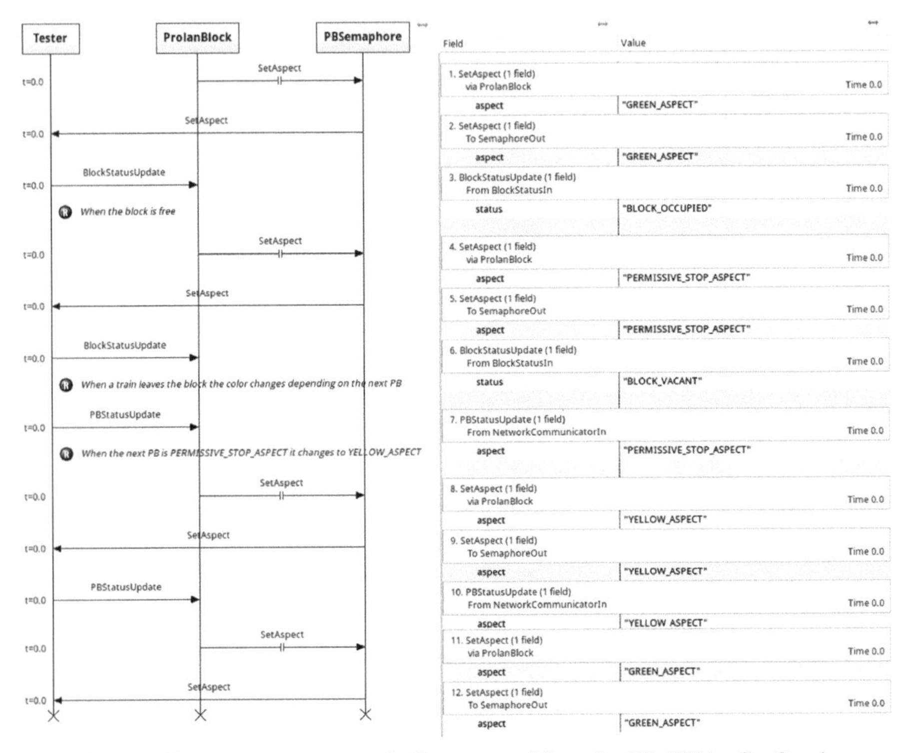

Figure 2.11 A test case automatically generated from the BB-PIT by Conformiq.

focus on a functional V&V modeling in the Integration Verification Design, whereas they focus on functional and structural V&V modeling at this stage.

Prolan assessed the IBM Rhapsody Automatic Test Generator (ATG) [36] for the structural verification of the *ProlanBlockCoreLogic*. ATG generated ten test cases (Figure 2.11) achieving the 91% coverage of the structural features of the model (they covered 19/21 states and 22/24 transitions). However, it was not able to reach the complete coverage.

2.4.8 Model-Driven V&V Subprocess

The activities on the right side of the V-Model concern V&V execution: we propose to use in these phases the models built during the Design activities, but after they have been refined with the new details added in PSM during implementation.

Thus, in Component Verification we named the model White Box Platform Specific Model (WB-PST), and the Tester adds new test cases considering the details of the target platform. The WB-PST can be used to calculate the test coverage on the basis of the final system code, as well as to support any kind of verification of the actual code, such as to derive consistent and efficient code review plans by considering the component software metrics and implementation details.

Similarly, during Integration Verification phase the BB-PIT model is refined in the Black Box Platform Specific Test Model (BB-PST), where platform specific details complete the integration test specification. For instance, the BB-PST can be exploited to perform interface testing, a technique that is highly recommended by CENELEC EN 50128: interface testing is executed knowing the actual domain of all interface variables, and selecting particular input to assess the behavior of the (integrated) components (e.g., at their normal, boundary, or invalid values).

Finally, in Validation phase, the Tester assesses that system and software requirements are met. To this end, (s)he executes the overall system tests defined in the CIT. Moreover, if the CIT is executable, the Tester can put the CIT and the software in-a-loop, to perform software-in-the-loop and hardware-in-the-loop testing.

2.5 Environment System Validation

Our methodology also exploits MDE for validation, by defining an environmental model during Validation Design to analyze the actors' behavior. This model, named CIT Model, specifies the expected behavior of the environment when interacting with the system by behavioral diagrams (e.g., Sequence, State Machine or Activity diagrams) that are used to derive validation test case.

The CIT also appears in other studies [37] but our definition differs from the previous ones, since we define the CIT a model for validation that abstracts from the computation details of the system under analysis (SUT), and also propose to develop the CIT as an executable model of the environment, with interfaces specular to those of the PIM. This definition supports multiple forms of V&V during the product life cycle.

Since the CIT has an interface specular to the one of the PIM, we can put the two models in-a-loop and the CIT can be used to perform model-in-the-loop test (since it is runnable), enabling to:

- Validate the system against its expected interactions with external actors;
- Create a simulated environment to reason about the operational aspects of the system in its environment (also through model animation).

Model-in-the-loop (MIL) testing can be performed as soon as the PIM is available, i.e., during the Component Design, enabling to an early fault detection. Moreover, if the Tester can use additional/external sources of knowledge to model the actors' behavior (such as domain knowledge or historical data), MIL testing can also be useful to detect missed software requirements, by assessing the behavior in a simulated environment.

Then, when a system implementation is available, the Tester can build an adapter to allow the CIT to interact with the actual SUT, allowing Software- and Hardware-in-the-loop Testing.

CIT also enables to performance testing, generally adopted for the assessment of critical systems, as it is recommended by the safety standards. Indeed, the CIT can generate inputs representative of the operational profile.

Other forms of verification allowed by the CIT are:

Model-checking through the in-the-loop model. This can assess the absence of undesired conditions during the operation, analyzing the states of the PIM and CIT.

Back-to-back testing, a special case is when the CIT can be seen as another PIM, for instance when we consider systems that act as client and server at the same time. For this kind of systems, we can instantiate two PIMs, putting in-a-loop each other, to perform back-to-back testing.

2.6 Experimenting the CIT

The benefits of the CIT and of environmental modeling have been assessed in another part of the interlocking system with which Prolan Block interacts, the Prolan Monitor.

The Prolan Monitor (PM) shares with the PB the Prosigma hardware and middleware platform, which is the basis of the next generation of Prolan's products.

The purpose of the PM is to send signals generated by legacy interlocking devices to modern interlocking systems that communicate through protocols based on IP networks (such as via X.25 over TCP/IP). More specifically, PM monitors *railway objects*: each object is associated to one bit of information, which is encoded by one couple of valent and antivalent physical signal values. The PM transmits the bit of information to other devices, detecting invalid values for the couple of electric signals. Indeed, the input can suffer of special unstable states during which the signals quickly alternate in their value for a transient time, called *bounce time*: the PM must properly filter the signals, separating transient noise from invalid inputs.

To assess the benefits of the CIT, Prolan made an executable model of the PM's environment. The CIT is composed of two *CIT Railway Objects*: each *CIT Railway Object* controls the couple of logical signals associated with the binary information that they encapsulate; from the CIT point of view, the PM is an actor.

The *CIT Railway Objects* are implemented by a *Signal Generator* and an *Event Generator*: the *Event Generator* determines the next output to be triggered (including transient and invalid states), as specified by a user-defined operational profile, whereas the *Signal Generator* sets the couple of output signals and manages the duration of the transients.

A panel diagram makes the CIT interactive: a couple of knobs allow to set the event generation period and to customize the duration of transient states.

Linking together with an adapter simulating a physical relay the CIT and the PIM, we preliminarily performed Model-in-the-loop testing. Only changing the adapter with a real hardware card forwarding the events to the actual SUT, we could also perform Hardware in-the-loop testing (Figure 2.12).

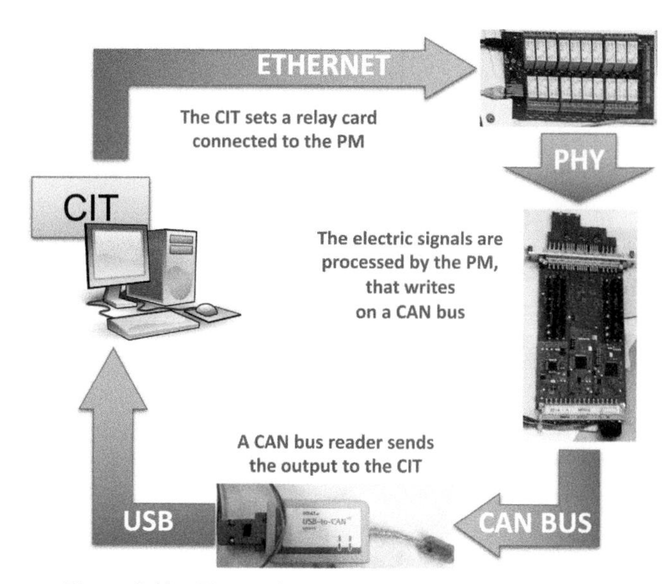

Figure 2.12 The configuration of the PM for HIL Testing.

2.7 Lesson Learned

The CECRIS knowledge transfer activities allowed assessing the maturity of MDE for railway interlocking systems. The project managers became acquainted with MDE methodologies and tools, and Prolan started to consider their introduction into the development processes.

Indeed, the pilot project showed that MDE is a mature technology, which supports the whole development process. Using SysML for requirements specification helped to produce better artifacts, reasoning formally on incongruences and missing specifications than with the current document-centric approach. Also, fast prototyping, early fault detection, automatic test generation, and other MDE features revealed a gain of productivity and quality during design and V&V phases than current methodology.

However, even for SME companies as Prolan that have limited engineering capacity, it is not easy to change current development process and practices. MDE requires for a technological and knowledge transfer. While the former can be addressed with personnel trainings, the latter is more subtle, long, and expensive. Therefore, Prolan submitted a joint tender proposal together with the Hungarian University, partner of CECRIS project, aiming at getting active support in their introduction during the next safety critical project, and to investigate model-driven technologies further.

Indeed, still an extensive experimentation of model-driven methodologies is needed. By the pilot project on the Prolan Block we qualitatively assessed MDE: even if the benefits of model-driven approaches turned out to be evident, we could not easily evaluate how much time is needed for Prolan to have a return of investment. This tender would introduce the academic know-how and support in the planning, design and implementation for a complete Railway Interlocking System project.

However, the current experience has been saved by Prolan, and if the tender is rejected then the enhancement of the current development lifecycle of Prolan with the one proposed in the framework of the pilot project will be applied. The innovation is planned to be applied gradually, through several stages expected to last several years.

At the first stage, Prolan targets to introduce models for supporting the current activities, starting to use MBE than MDE approaches. For instance, it is expected to adopt modelling tools for system requirement specification and system design phases. The current document-based system requirements specification and the the requirement management system will be replaced with software using SysML models, taking benefits of the improved model traceability.

These first changes already raise knowledge and technologies issues: we still have not decided if to adopt exclusively SysML for describing the system requirements, because we could be not able to teach satisfactorily modeling and SysML to all the team members; moreover, we have not completed the tool selection process. Among the selection criteria it is required that new tools be stable, and easily interoperable with the other software suites already in use at Prolan. The fully compliance with standards, and the vendor lock-in problem are also of interest for the tool selection. The relatively high price of licenses and the difficulty of using these modeling tools have an adverse effect on the introduction of model-driven methodologies.

The CECRIS experience revealed that model-driven technologies can really improve the development, enhancing quality of the product and of the development life cycle. Despite these advantages, the management of the railway product development decided to pursue a conservative approach towards the introduction of model based tools and development methods: the big issues of MDE concern skills and organization. The learning curve of these technologies is long and difficult to quantify, and the relevance of roles during the development will change, deeply impacting human-organizational factors. The innovation must be introduced gradually, taking into account these factors.

The benefits of cooperating with academic partners within the CECRIS project were manifold, as it was demonstrated by the pilot project: the experience in the methodologies and tools enabled Prolan to receive active tutoring and support for the full product lifecycle, shortening the learning curves and reducing the number of errors caused by the lack of knowledge, method, and experience with the tools. Moreover, the broad knowledge of the emerging new technologies in the field enabled the academic partners to suggest the criteria in selecting the appropriate tools and methodologies.

References

[1] Liebel, G., Marko, N., Tichy, M., Leitner, A., and Hansson J. (2014). "Assessing the State-of-Practice of Model-Based Engineering in the Embedded Systems Domain," in *Proceedings of the 7th International Conference on Model-Driven Engineering Languages and Systems (MODELS)*, eds. J. Dingel, W. Schulte, I. Ramos, S. Abrahão, and E. Insfran (Berlin: Springer International Publishing), 166–182.

[2] Marko, N., Liebel, G., Sauter, D., Lodwich, A., Tichy, M., Leitner, A., and Hansson J. (2014). Model-based engineering for embedded systems

in practice. Research Reports in Software Engineering and Management, Technical report, University of Gothenburg, Gothenburg.

[3] Broy, M., Kirstan, S., Krcmar, H., Schätz, B., and Zimmermann, J. (2013). "What is the benefit of a model-based design of embedded software systems in the car industry" in *Emerging Technologies for the Evolution and Maintenance of Software Models* (Hershey, PA: IGI Global), 343–369.

[4] CECRIS. (2016). *EU Project CECRIS, CErtification of CRItical Systems*. Available at: http://www.cecris-project.eu

[5] Schmidt, D. C. (2006). "Guest Editor's Introduction: Model-Driven Engineering," in: *Computer 39.2*, 25–31. Lecture Notes in Computer Science (Berlin: Springer).

[6] Brambilla, M., Cabot, J., and Wimmer, M. (2012). *Model-Driven Software Engineering in Practice*. San Rafael, CA: Morgan & Claypool Publishers.

[7] Object Management Group (OMG). (2014). *MDA Guide* (Version 2.0). Available at: http://www.omg.org/cgi-bin/doc?ormsc/14-06-01 (accessed on 2016-03).

[8] Baker, P., Dai, Z. R., Grabowski, J., Haugen, Ø., Schieferdecker, I., and Williams, C. (2007). *Model-Driven Testing: Using the UML Testing Profile* (New York, NY: Springer-Verlag New York, Inc.).

[9] Dai, Z. R. (2004). "Model-driven testing with UML 2.0," in *Proceedings of the 2nd European Workshop on Model Driven Architecture (MDA) with an emphasis on Methodologies and Transformations (EWMDA)*, eds D. Akehurst, 179–187. Tech. rep. 17-04, University of Kent, Canterbury.

[10] Kent, S. (2002). "Model Driven Engineering," in *the Proceedings of the Third International Conference on Integrated Formal Methods (IFM)*, 286–298. Berlin: Springer-Verlag.

[11] Object Management Group (OMG). (2003). *MDA Guide (Version 1.0.1)*. Available at: http://www.omg.org/cgi-bin/doc?omg/03-06-01 (accessed on 2016-03).

[12] Davies, I., Green, P., Rosemann, M., Indulska, M., and Gallo, S. (2006). How do practitioners use conceptual modeling in practice? *Data Knowl. Eng.* 58.3, 358–380.

[13] Forward, A. and Lethbridge, T. C. (2008). "Problems and Opportunities for Model-centric Versus Code-centric Software Development: A Survey of Software Professionals," in *Proc. of the International Workshop on Models in Software Engineering (MISE)* (New York, NY: ACM), 27–32.

[14] Hutchinson, J., Rouncefield, M., and Whittle, J. (2011). "Model-driven engineering practices in industry," in *Proceedings of the 33rd International Conference on Software Engineering (ICSE)* (New York, NY: IEEE), 633–642.

[15] Hutchinson, J., Whittle, J., Rouncefield, M., and Kristoffersen, S. (2011). "Empirical Assessment of MDE in Industry," in *Proceedings of the 33rd International Conference on Software Engineering (ICSE)* (New York, NY: ACM), 471–480.

[16] Hutchinson, J., Whittle, J., and Rouncefield, M. (2014). "Model-driven engineering practices in industry: social, organizational and managerial factors that lead to success or failure," in *Science of Computer Programming 89, Part B* (Amsterdam: Elsevier), 144–161.

[17] Whittle, J., Hutchinson, J., and Rouncefield, M. (2014). "The state of practice in model-driven engineering," in *IEEE Software 31.3* (New York, NY: IEEE), 79–85.

[18] Tomassetti, F., Torchiano, M., Tiso, A., Ricca, F., and Reggio, G. (2012). "Maturity of software modelling and model driven engineering: A survey in the Italian industry," in *Proceedings of the 16th International Conference on Evaluation Assessment in Software Engineering (EASE)* (New York, NY: ACM), pp. 91–100.

[19] Torchiano, M., Tomassetti, F., Ricca, F., Tiso, A., and Reggio, G. (2011). "Preliminary Findings from a Survey on the MD State of the Practice," in *Proceedings of the International Symposium on Empirical Software Engineering and Measurement (ESEM)* (New York, NY: ACM), 372–375.

[20] Torchiano, M., Tomassetti, F., Ricca, F., Tiso, A., and Reggio, G. (2012). "Benefits from modelling and MDD adoption: expectations and achievements," in *Proceedings of the 2nd International Workshop on Experiences and Empirical Studies in Software Modelling (EESSMod)* (New York, NY: ACM), 1–6.

[21] Torchiano, M., Tomassetti, F., Ricca, F., Tiso, A., and Reggio, G. (2013). Relevance, benefits, and problems of software modelling and model driven techniques – A survey in the Italian industry. *J. Syst. Softw.* 86.8, 2110–2126.

[22] Agner, L. T. W., Soares, I. W., Stadzisz, P. C., and Simo, J. M. (2013). A Brazilian survey on UML and model-driven practices for embedded software development. *J. Syst. Softw.* 86.4, 997–1005.

[23] Mussbacher, G., Amyot, D., Breu, R., Bruel, J. M., Cheng, B. H. C., Collet, P., Combemale, B., France, R. B., Heldal, R., Hill, J.,

Kienzle, J., Schöttle, M., Steimann, F., Stikkolorum, D., and Whittle, J. (2014). "The relevance of model-driven engineering thirty years from now," in *Proceedings of the 17th International Conference on Model-Driven Engineering Languages and Systems (MODELS)* eds. J. Dingel, W. Schulte, I. Ramos, S. Abrahão, and E. Insfran (New York, NY: Springer International Publishing), 183–200.

[24] Mohagheghi, P. and Dehlen, V. (2008). "Where Is the Proof? A Review of Experiences from Applying MDE in Industry," in *Proceedings of 4th European Conference on the Model Driven Architecture – Foundations and Applications (ECMDA-FA)*, Vol. 5095, ed. I. Schiefer-decker and A. Hartman. Lecture Notes in Computer Science. Springer Berlin Heidelberg, 2008, pp. 432–443.

[25] Baker, P., Loh, S., and Weil, F. (2005). "Model-Driven Engineering in a Large Industrial Context – Motorola Case Study," in *Proceedings of the 8th International Conference on Model Driven Engineering Languages and Systems (MODELS)*, eds. L. Briand and C. Williams (Berlin: Springer), 476–491.

[26] Weigert, T. and Weil, F. (2006). "Practical experiences in using model-driven engineering to develop trustworthy computing systems," in *Proc. of the IEEE International Conference on Sensor Networks, Ubiquitous, and Trustworthy Computing*, Vol. 1 (New York: IEEE), 208–215.

[27] Huhn, M. and Hungar, H. (2010). "8 UML for software safety and certification," in *Model-Based Engineering of Embedded Real-Time Systems: International Dagstuhl Workshop. Revised Selected Papers*, eds. H. Giese, G. Karsai, E. Lee, B. Rumpe, and B. Schätz (Berlin: Springer), 201–237.

[28] Pettit, R., Mezcciani, N., and Fant, J. (2014). "On the needs and challenges of model-based engineering for spaceflight software systems," in *Proceedings of the IEEE 17th International Symposium on Object/Component/Service-Oriented Real-Time Distributed Computing (ISORC)* (New York: IEEE), 25–31.

[29] Ferrari, A., Fantechi, A., and Gnesi, S. (2012). "Lessons learnt from the adoption of formal model-based development," in *Proc. of 4th International Symposium on the NASA Formal Methods (NFM)*, eds. A. E. Goodloe and S. Person (Berlin: Springer), 24–38.

[30] Object Management Group (OMG). (2008). Systems modeling language (SysML). Available at: http://www.omg.org/docs/formal/08-11-02.pdf (accessed on 2016-03).

[31] Scippacercola, F., Pietrantuono, R., Russo, S., Zentai, A. (2015). "Model-driven engineering of a railway interlocking system," in *Proceedings of the 3rd International Conference on Model-Driven Engineering and Software Development (MODELSWARD 2015)* (Setúbal: SCITEPRESS), 509–519.

[32] No Magic, Inc. (2016). *Magic Draw. MagicDraw*. Available at: http://www.nomagic.com/products/magic-draw.html (accessed on 2016-03).

[33] IBM Corp. (2016). *Rational® Rhapsody® Developer*. Available at: http://www-03.ibm.com/software/products/it/ratirhap (accessed on 2016-03).

[34] Conformiq Inc. Conformiq Designer. http://www.conformiq.com/products/conformiq-designer, (accessed on 2016-03).

[35] IBM Corp. (2016). *Rational® Rhapsody® Test Conductor Add On*. User Guide. Available at: http://pic.dhe.ibm.com/infocenter/rhaphlp/v7r6/topic/com.ibm.rhp.oem.pdf.doc/pdf/RTCUserGuide.pdf (accessed on 2016-03).

[36] IBM Corp. (2016). *Rational® Rhapsody® Automatic Test Generator Add On*. User Guide. Available at: http://pic.dhe.ibm.com/infocenter/rhaphlp/v7r5/topic/com.ibm.rhapsody.oem.pdf.doc/pdf/ATG-UserGuide.pdf (accessed on 2016-03).

[37] Schieferdecker, I. (2005). "The UML 2.0 Test Profile as a Basis for Integrated System and Test Development," in *Proceedings of Köllen Druck+Verlag GmbH, Jahrestagung der Gesellschaft für Informatik* (Germany: Köllen Druck & Verlag GmbH), Vol. 35, pp. 395–399.

3

SYSML-UML Like Modeling Environment Based on Google Blockly Customization

Arun Babu Puthuparambil[1], Francesco Brancati[2], Andrea Bondavalli[3,4] and Andrea Ceccarelli[3,4]

[1]Robert Bosch Center for Cyber Physical Systems, Indian Institute of Science, Bangalore, India
[2]Resiltech s.r.l., Pontedera (PI), Italy
[3]Department of Mathematics and Informatics, University of Florence, Florence, Italy
[4]CINI-Consorzio Interuniversitario Nazionale per l'Informatica-University of Florence, Florence, Italy

3.1 Introduction

In industries, it is often observed that system designers may not be CS/OO/SysML experts and often required lot of training and support to use the modeling tools. Ideally, designers should spend all their effort on modeling and nothing else. However, existing modeling tools have lot of issues related to installation and plug-ins.

The use of Google Blockly was envisaged for use of modeling and simulation of systems. Blockly is a visual programming library, used to model/program using interlocked blocks (Figure 3.1). Each of the blocks also support traditional input widgets such as labels, images, textbox, checkbox, combo box, etc. It can be configured in such a way that only compatible blocks can be connected together (i.e. can be made "valid by design"). Blockly supports code and XML generation, and requires only a modern web browser which can be run on any device or operating system.

However, Blockly was not readily useable for modeling SysML/UML like models. A lot of changes and customizations were made in Blockly to make it more suitable for such type of modeling.

Figure 3.1 Various types of blocks in Blockly.[1]

3.1.1 Goal

To create a tool to create object diagrams based on a UML/SysML profile, which is simple, intuitive, fast, and reduce cognitive complexity. Also, the tool must support rapid modeling and code-generation. On a whole, the goal was to design a tool to model, validate, query, and support simulation.

3.1.2 Blockly Customization

Below is the list of customization performed on Blockly to make it more suitable to create SysML/UML like models. (i) support constraints; (ii) support behaviors; (iii) support links; (iv) support viewpoints; (v) support intuitive maximize, collapse, and semi-collapse; (vi) support requirements management; (vii) guide user to select compatible blocks; (viii) Blockly to PlantUML conversion; (ix) Blockly to Python code generation; (x) Blockly to graph conversion and graph querying; (xi) support sequence diagrams; (xii) custom minification of JavaScript for faster loading; and (xiii) support cardinality and singleton blocks.

3.1.3 Model Transformation

A SysML/UML profile can be given as an input to the tool, which will be converted to an intermediately format in PlantUML. As PlantUML is a simple textual language,[2] conversion to PlantUML makes it easier to debug model transformation. This also makes possible to edit and add any extra features/statements in the converted PlantUML by hand/tool if the earlier format did not support certain features.

[1] https://blockly-games.appspot.com/puzzle

[2] http://plantuml.com/PlantUML_Language_Reference_Guide.pdf

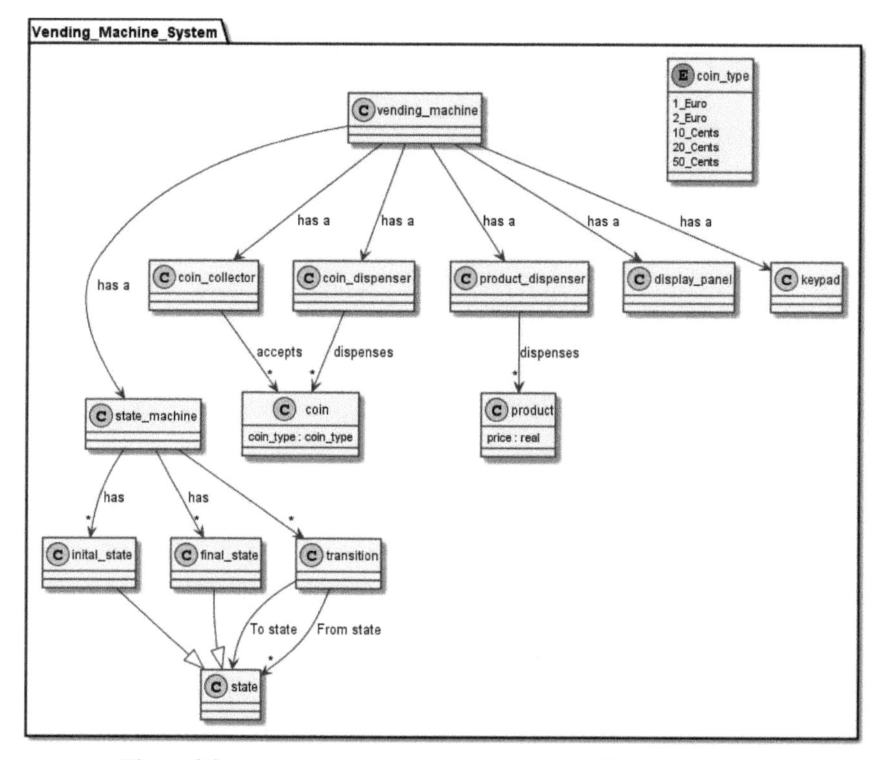

Figure 3.2 An example of a vending machine profile in PlantUML.

Figures 3.2 and 3.3 show a simple example of a vending machine system. The profile in PlantUML is given as input and the tool transforms it into interconnectable blocks.

3.1.4 Requirements Management

In the tool, each block can satisfy a set of requirements and a requirement can be satisfied by a set of blocks (Figure 3.4).

3.1.5 MDE Flow

The model-driven engineering (MDE) flow with the tool is shown in Figure 3.5. First, a profile expert provides a domain specific profile in SysML/UML. This profile restricts what a designer can design and which blocks are compatible with each other. The profile is then converted automatically to PlantUML and is imported into the Blockly format.

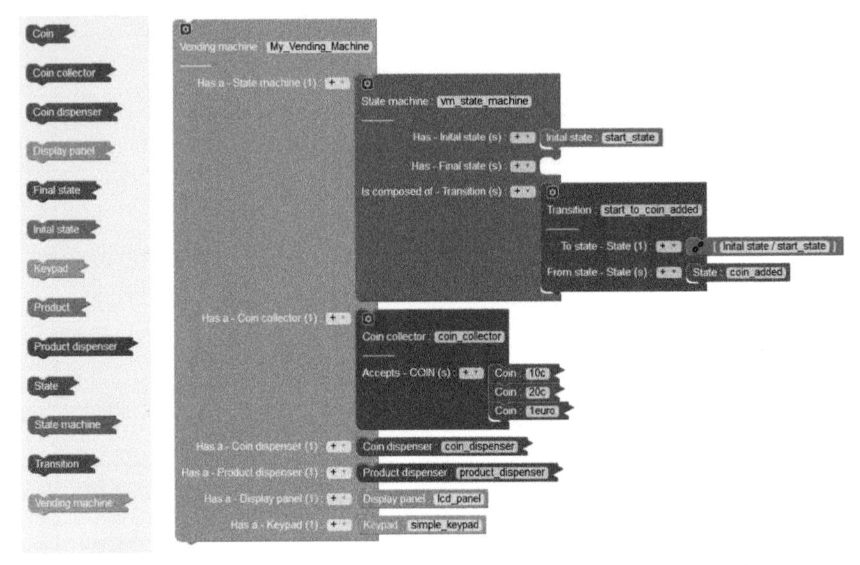

Figure 3.3 An exampleof a vending machine model under construction.

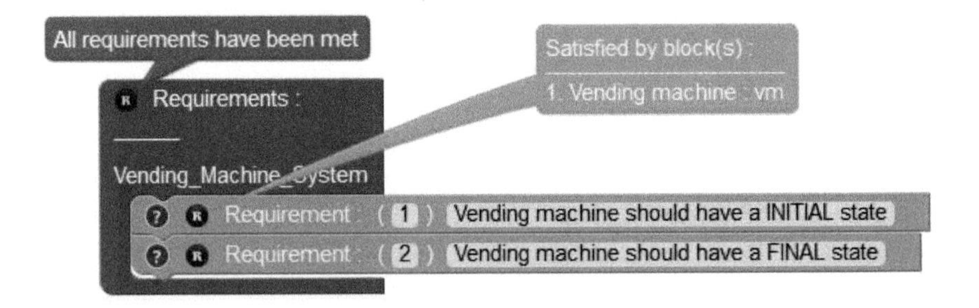

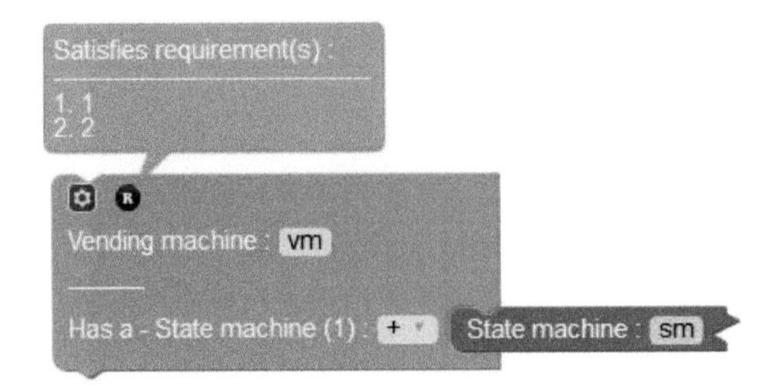

Figure 3.4 An example of requirements management.

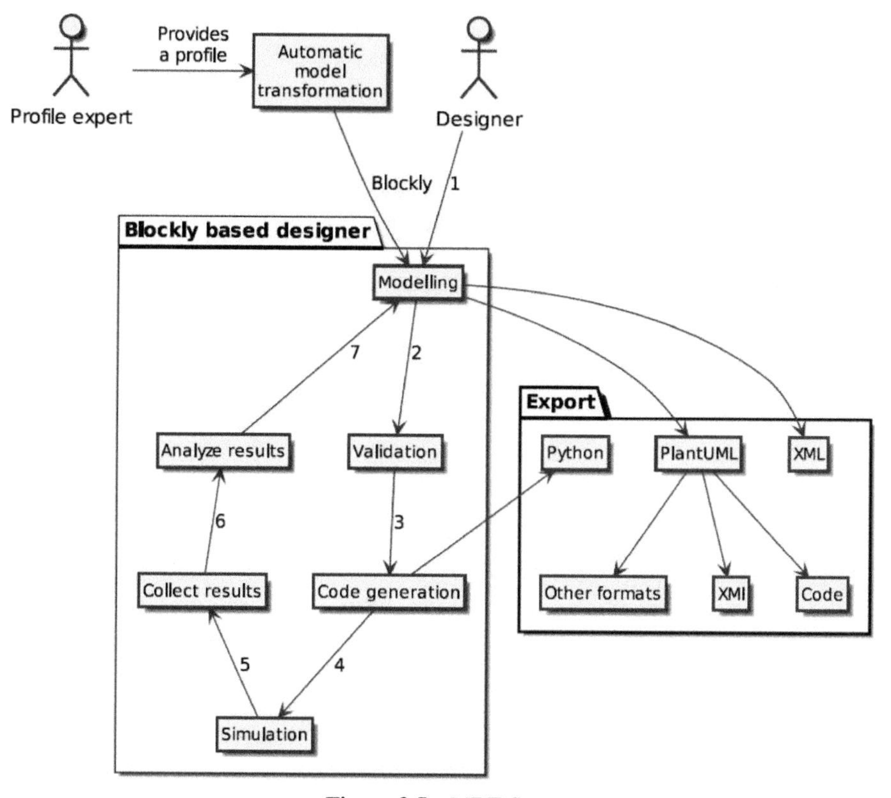

Figure 3.5 MDE flow.

The designer now uses the tool to model and validate the design. After validation code can be generated in Python programming language, which can be used for simulation/testing. After testing the results can be analyzed and changes can be made in the model if necessary. This cycle continues till the model is refined as necessary.

3.1.6 Guiding and Warning Users

The tool guides the designer in two ways: (i) Suggestions for the list of compatible blocks (Figure 3.6); and (ii) Using the type-Indicator plugin[3] (Figure 3.7).

[3]https://github.com/SPE-Systemhaus/blockly-type-indicator/wiki/Type-Indicator

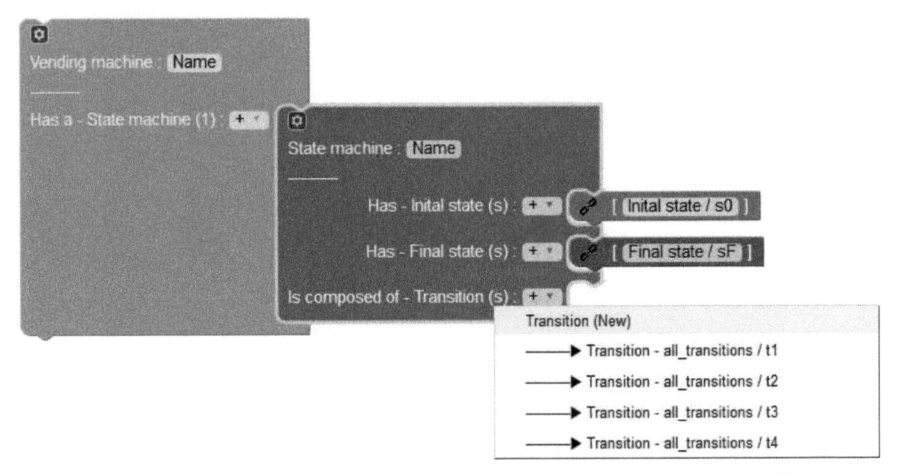

Figure 3.6 An Example of guiding users with compatible blocks (for Transitions).

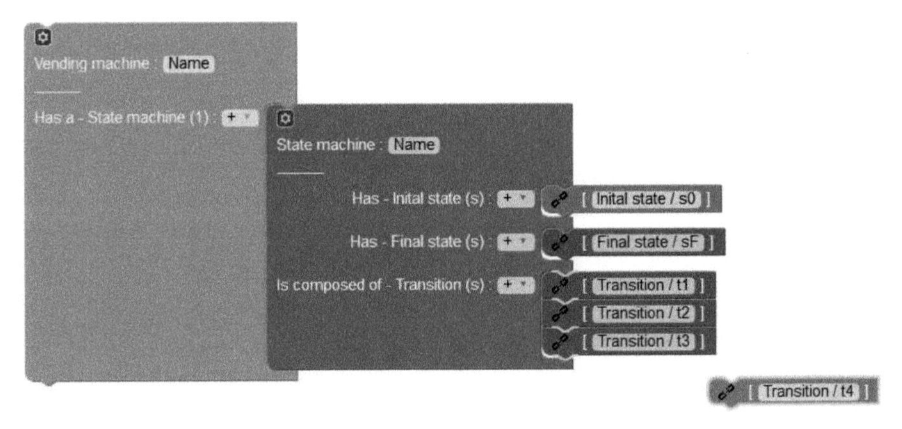

Figure 3.7 An example of type indicator plugin (Shows which blocks are compatible with the current selected block "Transition/t4" with yellow color).

Constraints make a model more precise, hence design time constraints are supported to warn designers when they make mistakes. These constraints are written in JavaScript and are evaluated at every *on change* event of a block (Figure 3.8).

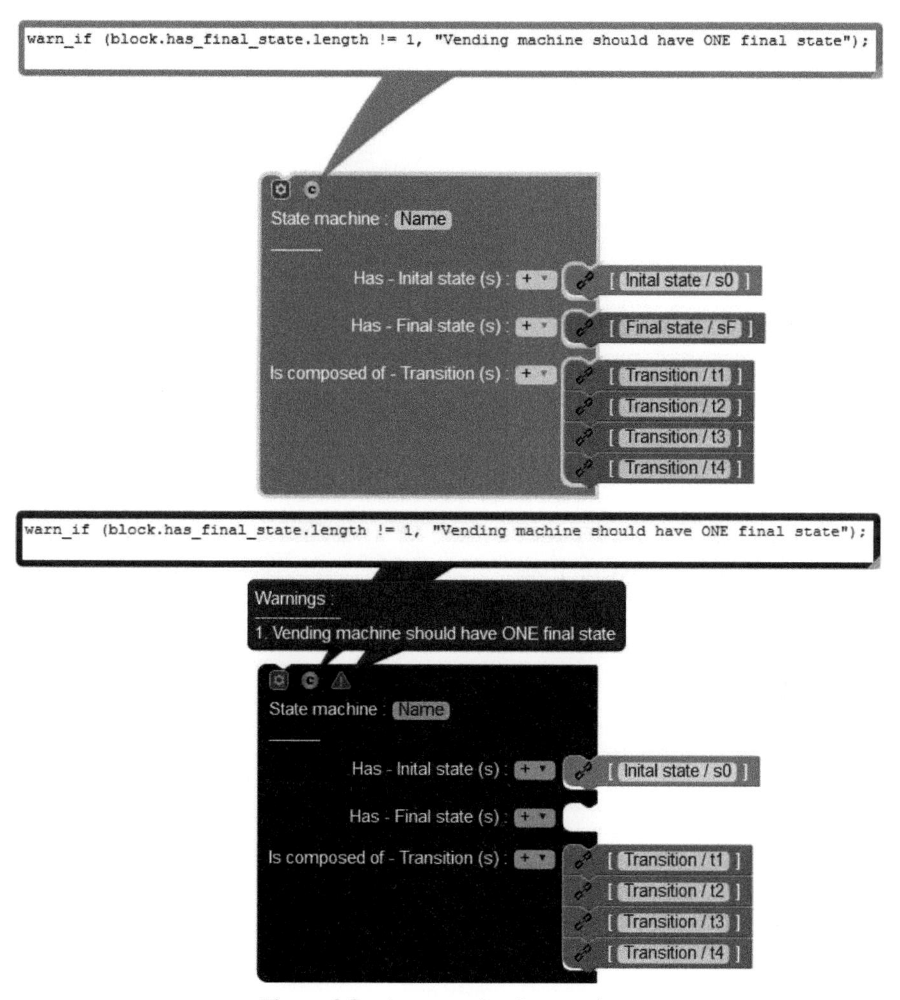

Figure 3.8 An example of constraints.

3.1.7 Modular Design and Viewpoints

Meaningful groups can be formed to modularize design and links can be used instead of lines to connect two blocks which are away from each other. Use of groups and links avoid the spaghetti diagrams in large models (see Figure 3.9).

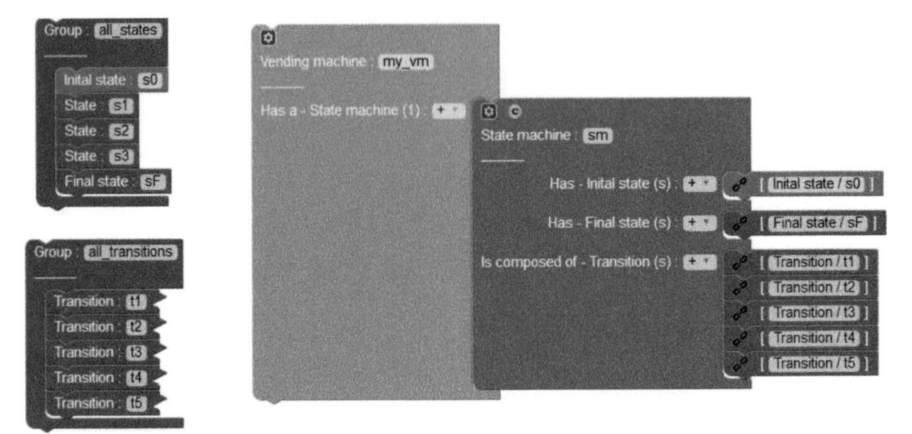

Figure 3.9 An example of groups and links.

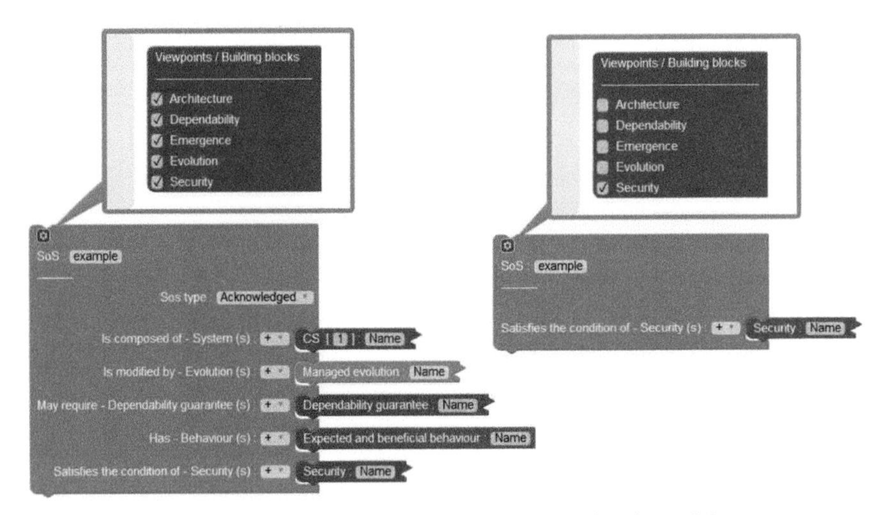

Figure 3.10 Enabling and disabling viewpoints in model.

Viewpoints are used in profile to reduce cognitive complexity for the designers. Viewpoints allow users to focus on one aspect of the model, e.g.: Architecture/Communication etc. Usually viewpoints do not exist in isolation; various viewpoints have relationships between each other. Figure 3.10 is an example of viewpoints in the tool; the viewpoints can be enabled/disabled.

3.1.8 Model Querying

On large models, it is important to query for blocks satisfying certain conditions. Thus, support for model querying in JavaScript was provided in the tool. The user provides a filter function, which is checked with all blocks. If the condition is satisfied, then it is highlighted, else it is not. Figure 3.11 is an example of the query "return true;" query, i.e., does not apply any filter and show all blocks. In Figure 3.12, instead, a filter is applied: it selects all blocks of type "RUMI" (return block.of_type == 'RUMI').

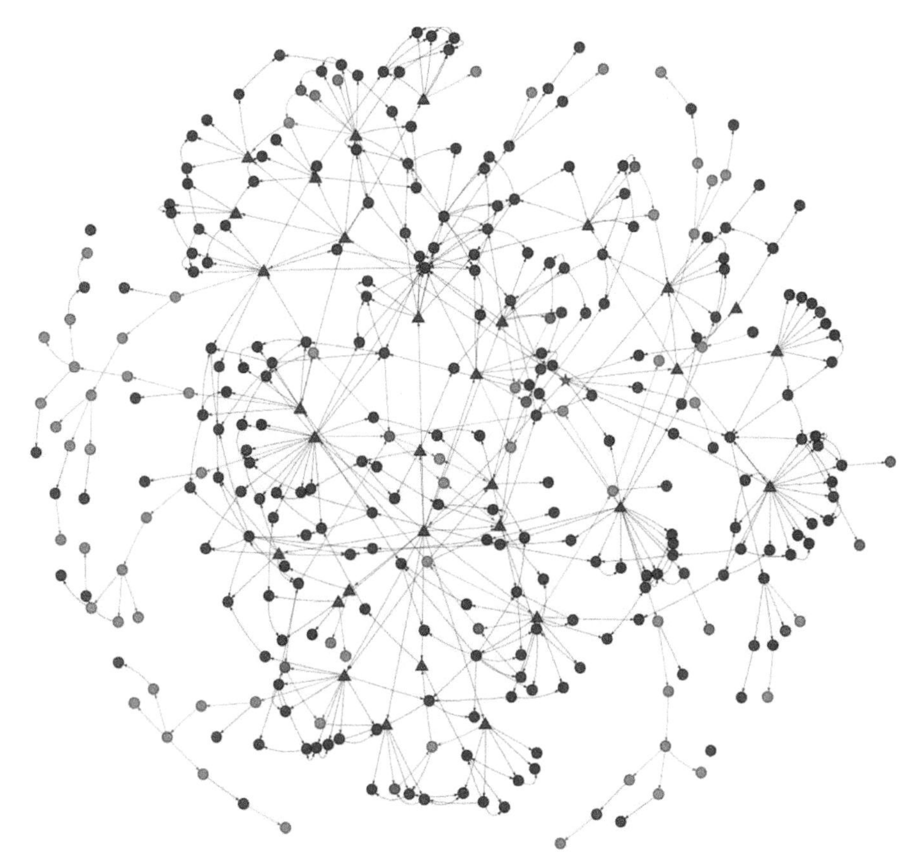

Figure 3.11 Model query without any filter (return true;).

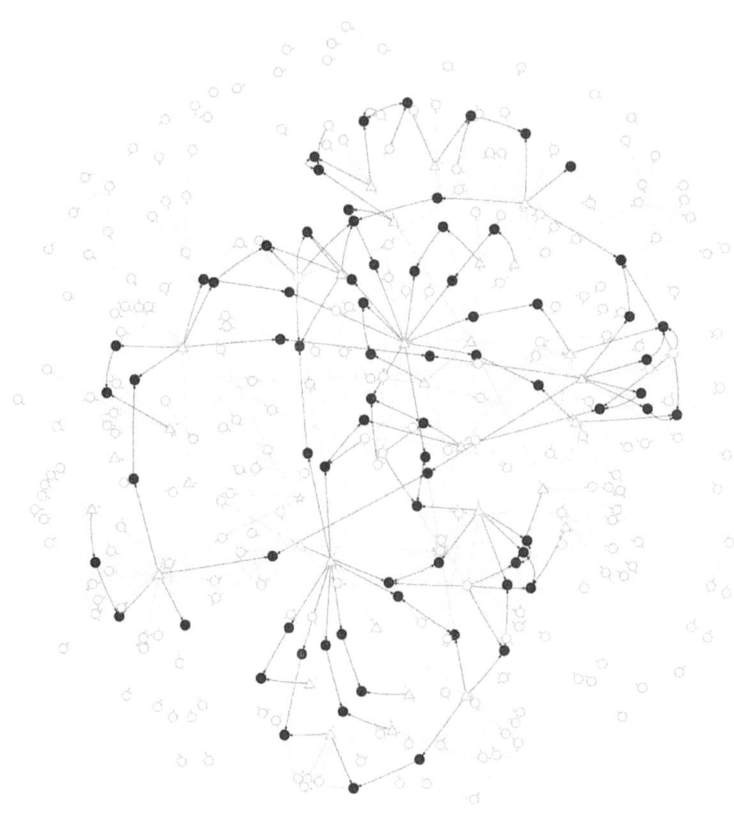

Figure 3.12 Example of model query to select all blocks of type "RUMI" (return block.of_type == 'RUMI').

3.1.9 Code Generation and Export to PlantUML

From the model, code can be generated to Python automatically. Python was chosen as it is one of the simplest object oriented programming language. However, other programming languages can easily be supported in Blockly.[4]

Also, as the model is available in .xml format and PlantUML format, custom code[5] and other programming languages can be supported in future.

Blockly models can be exported to PlantUML (Figure 3.13). The PlantUML version of model consist of two type of diagrams (i) the whole

[4]https://developers.google.com/blockly/guides/configure/web/code-generators

[5]https://developers.google.com/blockly/guides/create-custom-blocks/generating-code

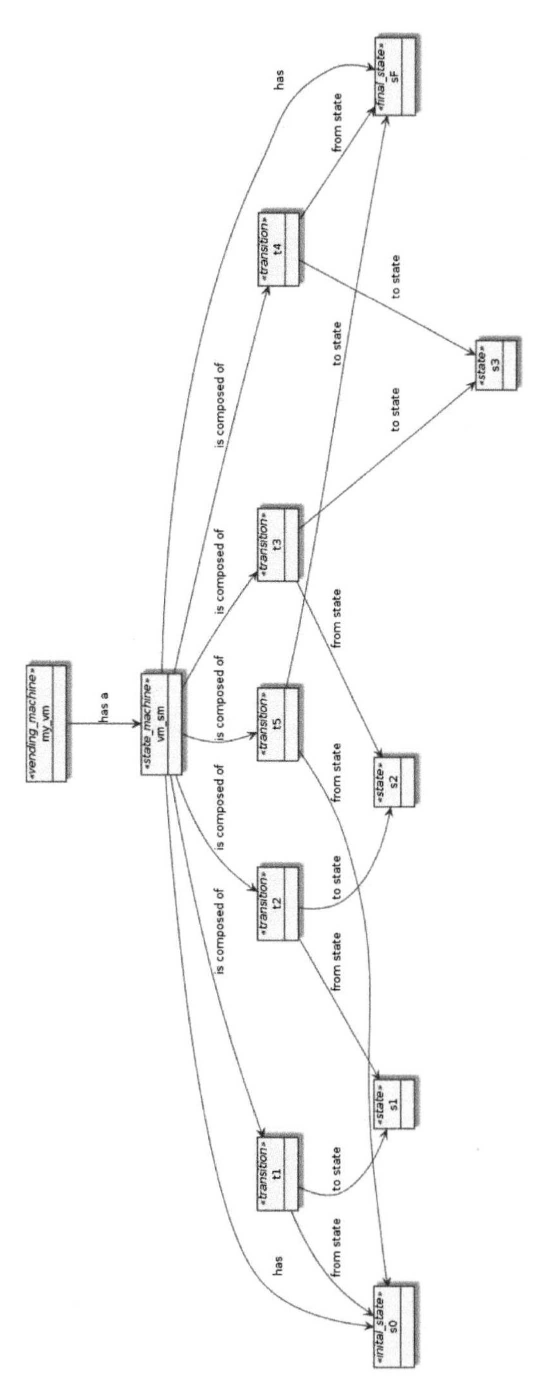

Figure 3.13 The subset of example model of "Vending machine" exported to PlantUML.

model without viewpoints; and (ii) model divided into separate files with grouped as viewpoints. These PlantUML models can be used for further refinement or be used with other tools.[6]

3.1.10 Simulation

Simulation of scenarios is supported using sequence diagrams and simulation related blocks custom code in Python. Domain specific sequence diagrams blocks are supportedto make design easier and error-free. As opposed to traditional generic sequence diagrams, these domain specific blocks are non-ambiguous and it allows correct code generation. Figure 3.14 shows an example of a sequence diagram containing domain specific blocks. Each sequence diagram can consist of sub-sequence, which in turn can consist of simple blocks such as: if, while, parallel, etc., and may also contain custom domain specific blocks.

The sequence diagram drawn using blocks can also be automatically converted to classical sequence diagram view as shown in Figure 3.15.

Custom code to be run before starting and after ending simulation can also be added using the simulation related blocks (Figure 3.16). These blocks can be used in initializing variables before simulation, pre-processing of data before simulation, and post-processing of results after simulation.

3.1.11 Conclusion and Future Work

This chapter has introduced an intuitive and simple semi-formal tool to be used to model, validate, query and simulate systems based on a SysML/UML profile. There is always scope to improve upon the approaches proposed in the chapter especially to make semi-formal methods popular among non-experts.

Some of the possible future research areas are: (i) Blocks with images or blocks shaped as images could make a great feature to make the design more intuitive (Figure 3.17); (ii) Currently, a transformation from Eclipse/Papyrus to PlantUML is available and can be readily used by the tool, however many more transformations can be written to PlantUML; and (iii) More programming languages support could be added in future.

---

[6]http://plantuml.com/running

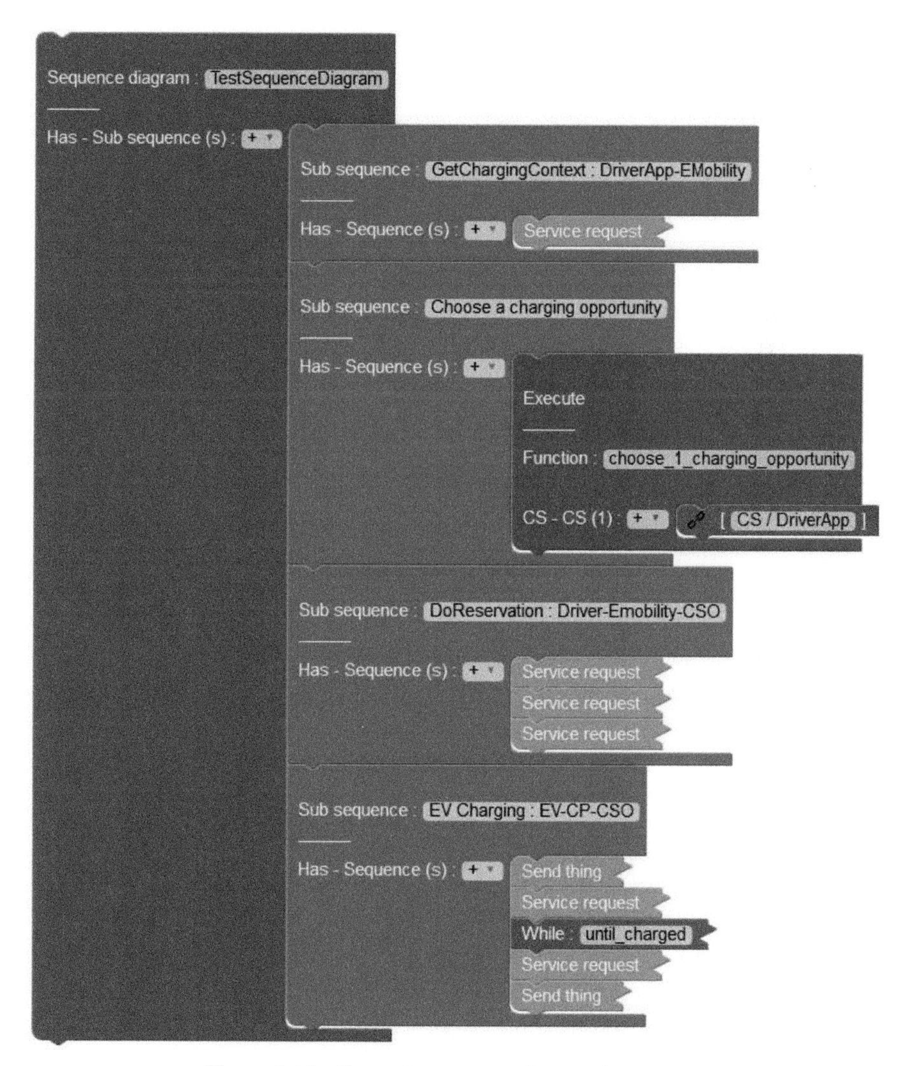

Figure 3.14 Example sequence diagram in Blockly.

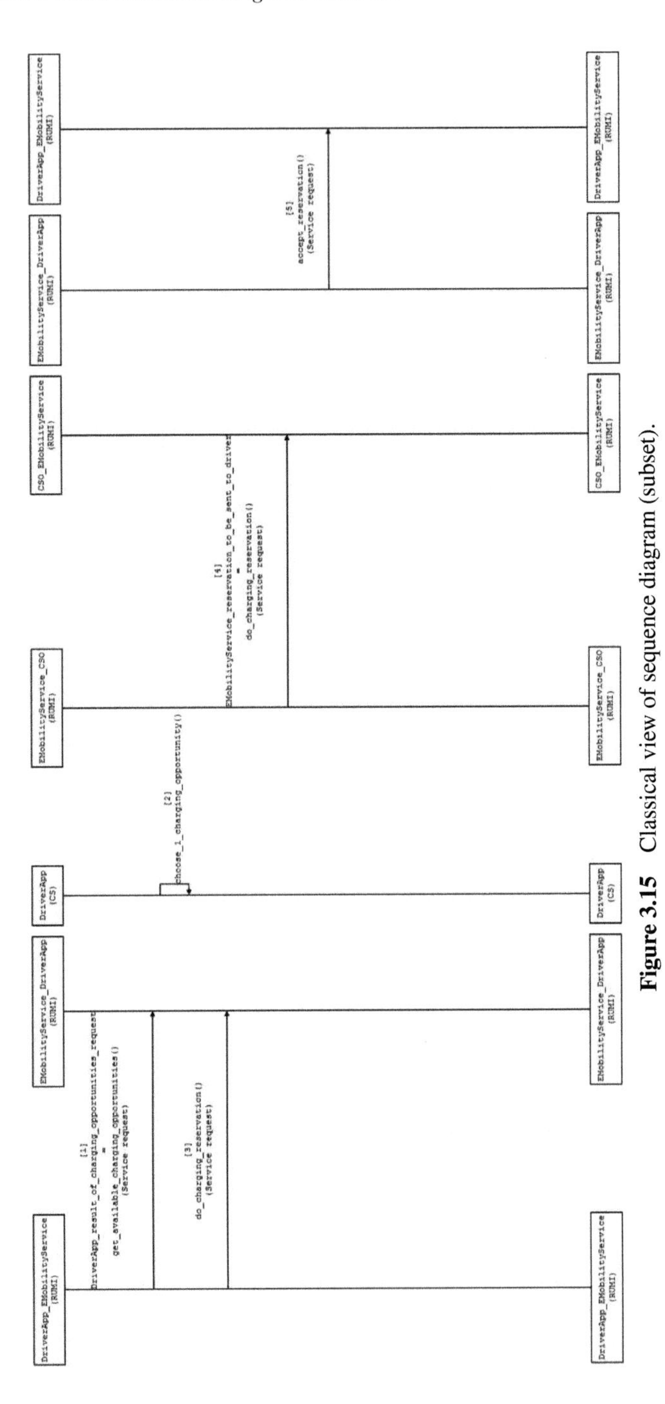

Figure 3.15 Classical view of sequence diagram (subset).

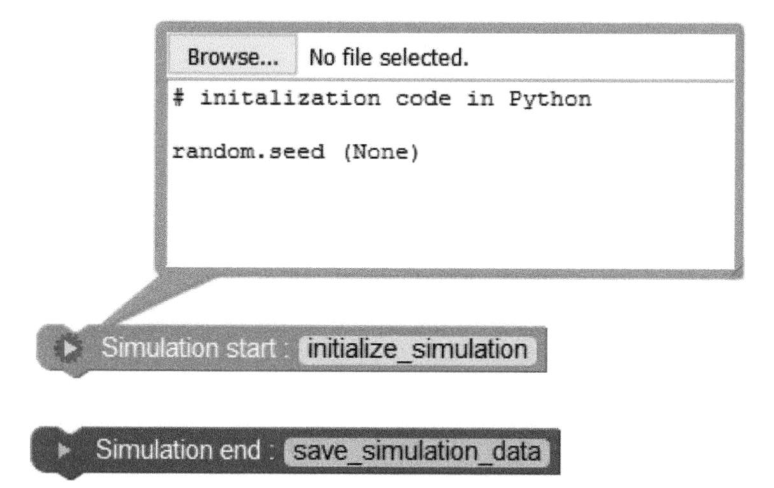

Figure 3.16 Blocks to support custom simulation initialization and code to execute when simulation ends.

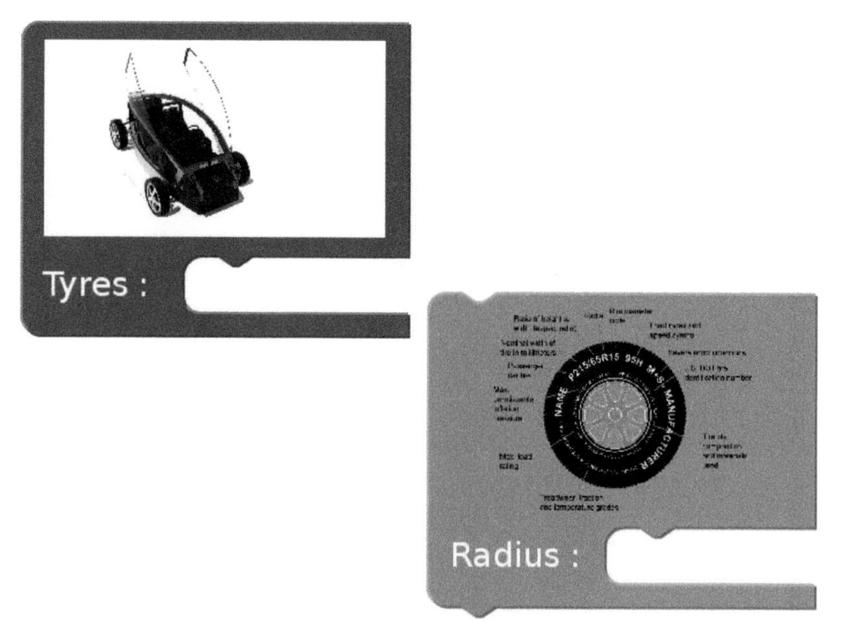

Figure 3.17 Blocks with images.

4

A Process for Finding and Tackling the Main Root Causes that Affect Critical Systems Quality

Nuno Silva[1], Francisco Moreira[1], João Carlos Cunha[2,3] and Marco Vieira[3]

[1]CRITICAL Software S.A., Coimbra, Portugal
[2]ISEC – Coimbra Institute of Engineering, Polytechnic Institute of Coimbra, Portugal
[3]CISUC, Department of Informatics Engineering, University of Coimbra, Portugal

4.1 Introduction

Following standards and applying good engineering practices during software development is not enough to guarantee defects free software, thus additional processes, such as Independent Software Verification and Validation (ISVV), are required in critical projects. The objective of ISVV is to provide complementary and independent assessments of the software artifacts in order to find residual defects and allow their correction in a timely manner. Independence is the most important concept of ISVV and it has been referred to and used in safety-critical domains such as civil aviation (DO-178B [1]), railway signalling systems (CENELEC [2]), and space missions (European Cooperation for Space Standardization (ECSS), e.g., [3, 4]). However, such systems are still far from being perfect and it is common to hear about software bugs in aeronautics, train accidents caused by software problems, satellite systems that need to be patched after launch, and so on.

Previous studies have analysed the results of ISVV activities [5, 7], looked into consolidated ISVV metrics [8] and studied the importance of independent test verification [9], showing that existing standards and good engineering practices are not enough to guarantee the required levels of safety and dependability of critical systems (CSs). Independence of Verification and Validation (V&V) avoids author bias and is often more effective at finding defects and failures. Independence can be managerial, financial or technical, brings separation of concerns, complementarity, second/alternative opinions, and also has the merit of pushing development and in-house V&V teams to focus on the quality of their work. The role of independence at early development phases is highlighted in EasterBrook [10] and clearly stated in the requirements of several standards such as CENELEC [2] (depending on the SIL level), and DO-178 [1] (where, for example, for the most critical level – Level A – 33 out of the 71 objectives/requirements of the standard must be satisfied with full independence).

The Orthogonal Defect Classification (ODC) [11] is a generic classification technique that turns semantic information in the software defect stream into a measurement on the process where defects have been caused, enabling an efficient root cause analysis. ODC [11] can be applied to the defects identified during ISVV in order to study their classifications (namely: type, the fix that removed the defect; trigger, the defect identification activity/condition; and impact, the effect of the defect if not corrected). ODC is the most commonly used defect classification scheme, but it was not specifically developed for CSs, or for systems that need to fulfil specific certification requirements.

The application of ODC to the defects identified during ISVV has been described in Silva and Vieira [12]. In that work, we used ODC to classify a dataset of 1070 development, validation, and operation defects from space applications that followed ECSS standards. The conclusions were that most of the defect types found are related to: (i) documentation issues (this is logical since the ECSS processes are heavily based on documentation evidences); (ii) functionality issues (generally related to requirements understanding and source code bugs that compromise the foreseen functionalities); and (iii) defective implementations of the planned functions (algorithms). The classification has also shown that the main defect triggers are related to document consistency, traceability activities, and test activities. Also, the main impacts include system capability, reliability, maintainability, and documentation quality. However, the key conclusion is that a large number of issues could not be classified due to unfit taxonomy of defect types, triggers and impacts, causing many doubts in the classifications (more that 30% of the cases).

In order to enhance ODC for better applicability to CSs, thus covering all ISVV defects and easing the classification for industry, as in Silva and Vieira [13, 29], we proposed specific adaptations of the taxonomies of three classification attributes: Type, Trigger, and Impact. The enhanced classification enabled the full coverage of the defects in the dataset, providing more precision and a sounder root cause analysis support. The adaptation has been defined after conducting the classification of the 1070 defects with the original ODC (presented in Silva and Vieira [12]) and by carefully analysing the classification gaps. To validate the modifications, this enhanced version of ODC was used to reclassify the entire dataset, allowing its full classification. However, the work presented in Silva and Vieira [13] does not concretely contribute to understanding the problems that lead to the defects, which motivates the root cause analysis and the suggestions for improvements performed in the present work. The work described in this chapter represents the definition of a defects assessment process and the results of the application of this process to a space systems defects dataset.

This chapter presents an analysis on trends, common (and uncommon) problems and their causes, and look at the general picture of critical defects within the software development lifecycle of space systems, considering our dataset of 1070 defects. The results are intended to help engineers in tackling the problems starting from the most frequent ones, instead of dealing with them one by one, as is traditionally done in industry nowadays. In practice, this work brings to light the main root causes of issues in space projects, which were identified, based on the defects classification and on relevant expert knowledge about those defects and about the software development process, contributing toward proposing improvements to the processes, methodologies, tools, standards, and industry culture.

The ultimate objective of this work is to enable, through a proposed assessment process, a detailed analysis of the defects and identification of their sources (common root causes) in order to: (i) avoid their introduction (by tackling the main deficiencies in software engineering); and (ii) allow a more efficient detection of the remaining defects during the software development lifecycle (by identifying appropriate V&V methods and techniques). To support our work, the results of the enhanced ODC taxonomy proposed in Silva and Vieira [13] are used as input and analysed in detail to support the root cause analysis.

4.2 Background

This section presents some background concepts, namely in what concerns the Orthogonal Defect Classification (ODC), ISVV, and previous relevant works.

4.2.1 Orthogonal Defect Classification

The ODC, originally proposed by IBM (Chillarege et al. [11]), is one of the most used defects classification approaches. It is intended to be generic and applicable to different technology domains, but it is mostly oriented to design, code and testing defects. ODC defines eight attributes for defects classification, divided into two main groups: (i) opener, and (ii) closer. Three attributes (Activity, Trigger, and Impact) classify the defect when it has been discovered and so they are part of the opener group. The other five attributes (Target, Type, Qualifier, Age, and Source) are used when the defect is resolved, being thus part of the closer group. The full taxonomies for each attribute can be obtained from the ODC v5.2 specification and are not included here for brevity. Nevertheless, a description of ODC attributes is summarized in Table 4.1.

In addition to ODC, several other classification taxonomies exist, including Beizer's [14], and IEEE Standard Classification for Software Anomalies [15]. Although ODC comprises some questionable attributes (8 dimensions), making it also somehow complex to classify, we have selected this taxonomy due to its generic nature, its orthogonality, its comprehensiveness and the level of usage in industry that seemed higher than for all the others. Also, it is important to emphasize that ODC has been used in the past as a starting point for developing new and focused defect taxonomies

Table 4.1 Orthogonal defect classification attributes description

ODC Attribute	Description
Activity	The actual activity that was being performed at the time the defect was discovered. The main activities applicable to this work are: Requirements verification, design verification, code verification, test verification and test execution.
Trigger	A trigger represents the environment or condition that had to exist for the defect to surface.
Impact	The impact is the effect that the team who is classifying the defect thinks it would have on the system if not corrected.
Target	Represents the high level identity of the entity that was fixed.
Type	The defect type is defined according to the fix that is necessary to remove it from the system. For that reason, it is best classified by a team/person who applied the fix to the defect.
Qualifier	Captures the element of a non-existent, wrong or irrelevant implementation.
Age	Categorizes the age of the defect, whether if it is new or surfaced from a previous defect.
Source	Describes the source of the defect in terms of its developmental history.

for different domains. A few examples were presented by Leszak et al. [16] and Lopes Margarido et al. [17], which used ODC for studying, building and validating defect categorization schemes. In practice, the focus of ODC is to support the analysis and feedback of defect data targeting quality issues from different phases of the engineering lifecycle.

4.2.2 Independent Software Verification and Validation (ISVV)

Independent Software Verification and Validation is a set of structured engineering activities and tools that allow independent analysts to evaluate the quality of the software engineering artifacts produced at each phase of the development lifecycle. ISVV is performed on mature artifacts, which follow a strict engineering standard and that have been previously verified and validated as part of the development process. It provides an additional layer of confidence and is not expected to find a large number of severe defects.

Independent Software Verification and Validation produces evidences that support measuring the quality of the software and related processes and is referenced in several international standards: (i) ISVV guide from the European Space Agency (ESA) [18]; (ii) ISO Software Lifecycle Processes (ISO/IEC 12207) [19]; and (iii) IEEE Software V&V (IEEE 1012) [20].

Independent Software Verification and Validation includes six phases (Figure 4.1) that can be executed sequentially or selected/adapted as the result of a tailoring process based on a criticality analysis [18].

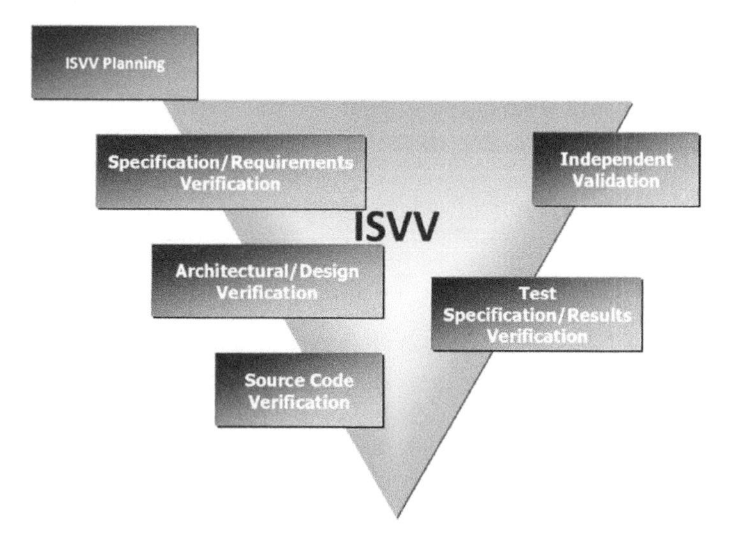

Figure 4.1 ISVV phases.

According to the ESA ISVV Guide [18], ISVV engineers classify defects considering three severity levels: (i) Major (defect with a significant impact in the system dependability, quality or safety); (ii) Minor (defect with a minimum impact on the artifacts quality but not in the end system); and (iii) Comment (an improvement suggestion). Each ISVV defect is also classified according to an ISVV defect type (e.g., External Consistency, Internal Consistency, Correctness, Technical Feasibility, Completeness, Readability, and Maintainability).

4.2.3 Related Work

Some studies in the literature have analysed metrics, efficiency and efficacy of the techniques used within ISVV to identify the defects in critical projects [5–8]. However, none of these studies considered their observations and results to classify the defects and improve the development processes, techniques, tools, or standards. Furthermore, we could not find in the research literature any complete study focused on defects in mission- and safety-critical systems, nor an extensive and complete classification or root cause analysis that relates the results of ISVV with the development lifecycle parameters of the systems under study. For space systems, Jones [21] has provided a small study about space failures in the frame of the European Space Agency missions, but simply concluded that the main cause for all the accidents was lack of testing. A more in-depth analysis is necessary as testing is not the cause but one of the detection methods.

Several researchers have looked into the analysis of failures in safety-critical systems during different life-cycle phases (from requirements to operations) and performed empirical studies and root cause analysis [22–24]. For example, Seaman et al. [25] used historical datasets with defects data from reviews and inspections and applied different categorization schemes to the defects. However, none of the mentioned studies covers all the life-cycle phases for the used defects dataset, nor bases the root cause analysis and the defects avoidance measures in a sound orthogonal classification of the defects.

Regarding the root cause analysis topic, it is worth mentioning some works that relate and somehow present results that are connected to the work presented in this chapter. Neufelder [26] collects data from field defects since 1993 and correlates that data to find the process properties that generate more defects; however, she is not focusing on CSs or systems developed under strict requirements and standards. Rao [27] has made an industry study about root cause defect classification for documentation defects, analysing only a few dozen defects on a monthly basis. Kumaresh et al. [28] conducted a study with data from a few hundreds of collected defects, where these defects have been classified and the corresponding root causes have been proposed to the learning of the projects as preventive ideas. No work has been performed for CSs nor with such a complete assessment and coverage of so many defect types (as shown by the ODC defect type results), as we did in our work.

4.3 Defects Assessment Process

Based on the analysis that we conducted and the lessons learned, we propose a general approach for root cause analysis of critical software, enabling the continuous improvement of implementation and V&V at all levels (processes, techniques, tools, personnel, application of standards, organization, and so on). Although our dataset and our experience are mainly from space software, we believe that this generalization is able to support the evaluation and root cause analysis of any critical system, independently from the domain. Figure 4.2 shows the general approach of a defects assessment procedure, which includes a root-cause analysis and a continuous improvement procedure, described hereafter.

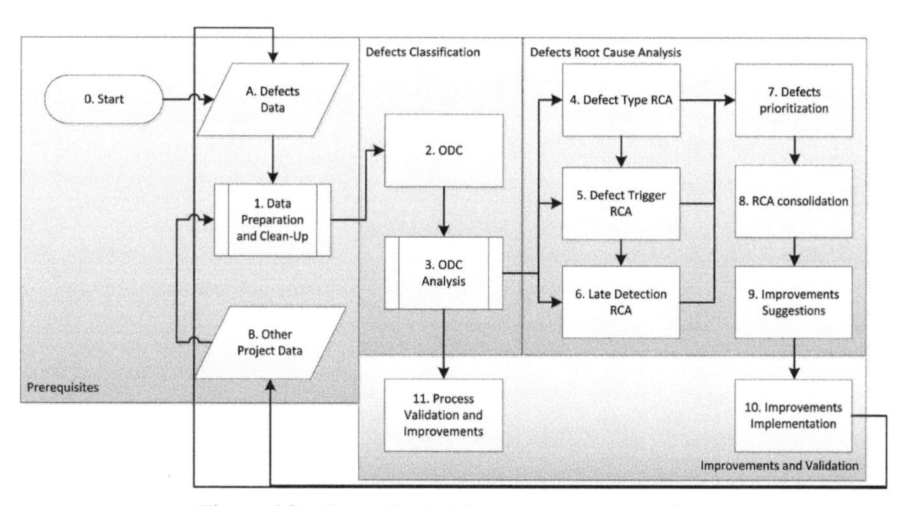

Figure 4.2 Generalized defect assessment procedure.

4.3.1 Procedure Prerequisites

The approach is based on data analysis and software engineering knowledge that require some prerequisites to be fulfilled for the correct application of the process:

0. Start:
In order to successfully perform the defects analysis, it is necessary that the collected data (A. Defects Data and B. Other Project Data) contain the necessary information. This includes basic requirements such as: (i) detailed information about each defect and its fix; (ii) knowledge of defect environment conditions, such as tools, personnel and constraints; (iii) engineers' assessment of the defect causes; and (iv) phase when the defect was introduced and when it was detected.

Some prerequisites are necessary to successfully apply the process. The first one includes training on the involved techniques, such as defects classification (e.g. ODC) and root cause analysis. The second includes rules and guidelines (or a template) for defects description or defect data collection.

1. Data preparation and clean-up:
Once we have the necessary data it is important to organize it and perform some anonymization if required. Data organization is essential for the next steps, since it is important to have the data in a searchable and manageable manner.

4.3.2 Defects Classification

In order to efficiently and concretely tackle the important problems of critical software engineering, the first set of activities shall focus on an orthogonal classification of the sets of defects:

2. ODC:
Perform the ODC classification on the organized dataset. Enhancements and adaptations to the ODC taxonomy can be useful depending on the nature of the defects and the domain; however, these enhancements should be quite precise. For examples, see Silva and Vieira [12, 13].

3. ODC Analysis:
Provide a summary of the ODC analysis. This information gives the first hints about the quality of the dataset, which can provide some feedback to the implementation and V&V teams. For examples, Silva and Vieira [12, 13].

4.3.3 Defects Root Cause Analysis

The root cause analysis is composed by several steps that include analysis of the defects types, the triggers allowing defect detection, the defects that could have been detected earlier, and then prioritization and consolidation of these root causes leading to concrete proposed improvements:

4. Defect Type RCA:
Based on the different defect types, identify the possible generic root causes. This list of causes shall come from experience and expert judgement or a dedicated database where defect types are mapped to root causes. The list of root causes might be reduced or harmonized in step (8) below.

5. Defect Trigger RCA:
Based on the classified defect triggers identify the causes and V&V techniques (or triggers) that allowed the defects detection at the current defect detection stage. This list of causes shall come from experience and expert judgement or a dedicated database where defect triggers are mapped to root causes. The list of root causes might be reduced or harmonized in step (8) below.

6. Late Detection RCA:
With the list of defects that have slipped more than one lifecycle phase milestone identify the causes of the failures in the V&V and ISVV techniques that allowed the defects to propagate until a later stage in the development lifecycle. This list of causes shall be added/harmonized with the list from step (5) Defect trigger RCA.

7. Defects prioritization:
If required (for example to tackle the defects with high impact on the system, or due to the large amount of defects and respective causes) list of defect types and triggers can be prioritized according to a defined severity (for example, based on the main impact of those defects) and the respective root causes can be filtered according to the prioritized type and trigger.

8. RCA consolidation:
The list of root causes obtained in the previous steps (4–6) is consolidated according to the prioritization done in step (7). This consolidation can also contribute to reduce to an essential and more concrete list of causes.

9. Improvements Suggestions:
For all the root causes, define solutions or modifications to the processes, techniques, tools, training, resources, environment or application of standards. The solutions must cover the development activities to avoid the creation of defects and also the defect detection activities in order to identify the defects as soon as possible.

4.3.4 Improvements and Validation

The suggested improvements might be difficult to implement, and their effectiveness can vary from team to team. They shall contribute to improve the software quality and reduce the amount of defects, different defects can then surface, and this is why this process shall have a consistent process improvement in place:

10. Improvements Implementation:
The development and V&V teams must be informed about the required changes or adjustments (9. Improvements Suggestions), and the organization, management and quality planning shall decide on the improvements to implement for future projects.

11. Process Validation and Improvements:
At every step, it is possible to derive improvements to the process. Such improvements can be set to adjust to the company culture, to the project environment, to the customer requirements, etc. However, it is essential to measure the effectiveness of the implementation of the results (9. Improvements Suggestions and 10. Improvements Implementation) once the suggestions have been implemented and new defects (or no defects) have been collected. Note that Improvement can and shall also be about the current process, the defects classification scheme, the root cause analysis techniques and so on. The presented process shall be able to adapt and help in improving itself and the related techniques that compose it.

4.4 Results

This section presents the dataset case studies description and the results of application of the process described in the Section 4.3.

4.4.1 Characterization of the Systems

Our analysis is based on a set of real defects from ISVV activities in space projects. The projects include subsystems that compose satellite systems for three different domains (i) scientific exploration; (ii) earth observation; and (iii) telecommunications; covering different types of software, such as start-up or boot software, on-board application software, command and control units, payload software, and attitude and orbit control units. The engineering processes used in the selected missions were driven by the ECSS standards, namely the space engineering standard E-ST-40 [3] and the quality standard Q-ST-80 [4] which has a comparable lifecycle and similar strict requirements imposed by the European Space Agency.

The subsystems were developed according to functional and non-functional requirements mandated from ECSS and mission specifics. They were characterized by the following needs/objectives, which are common to space CSs, that were collected from the ECCS standards [3, 4] and the corresponding engineering interpretations of the specification documents from several missions:

- No crash or hang shall happen at any time;
- No dynamic memory allocation is allowed;
- Communication – Telemetry (TM)/Telecommands (TC) – must always be possible between ground control and the satellite;
- The system must implement a Safe Mode (with basic communications, patch and dump functionalities);
- Most systems shall have a very simple and stable start-up software (also called boot software);
- There must be a watchdog (Hardware and/or Software) or an alive signal;
- Systems are built with redundancy (at least Hardware);
- Most systems must include FDIR (Fault Detection Isolation and Recovery) functionalities to account for the environment and external faults;
- The systems must have high autonomy and some self-correction procedures;
- Systems are categorized with a criticality level related to the impact or consequences of system failures (in this case, the ECSS defined levels are: Catastrophic, Critical, Major and Minor or Negligible).

The projects are also characterized by:

- Requirements written in natural language (structured), highly based on documentation and non-formal processes and languages;
- Documentation in UML/SysML and PDF, with limited possibilities of automated verification and formal analysis;
- Programming languages such as C, Ada and Assembly, that are quite mature and low level languages;
- Unit tests performed in commercial tools (e.g. Cantata++, VectorCast, LDRA), commonly developed and adapted for the specific projects embedded systems and environments;
- Integration and system testing performed in specific validation environment (Software Validation Facility – SVF) developed for this purpose on a case by case situation, with HW emulation and HW in-the-loop, simulated instruments, etc.

4.4.2 Defects in the Dataset

Table 4.2 summarizes the 1070 defects in the dataset, divided by severity (having a major or minor impact in the system, or just being comments to improve the engineering) and considering the ISVV activities in which they were found. The defects have been originated from the analysis of more than 10,000 software requirements, more than 1 million lines of code (mostly C, Ada95 and some Assembly), and over 3,000 tests[1] (some unit tests, some integration tests). In practice, the objective of ISVV was to find issues in the project artifacts, report and classify them in a clear and consistent way for the customer to act upon.

4.4.3 Enhanced ODC Results

The results of the application of the enhanced ODC for space defects are summarized in Table 4.2 showing the five top types, triggers and impacts cover about 90% of the issues analysed. This observation suggests that actions can be taken to quickly improve the quality of systems, by tackling a limited amount of properties.

[1]The 3,000 tests correspond to only part of the requirements and code referred, as not all ISVV activities cover the full set of artifacts, e.g., for some projects only source code analysis was performed, no tests related to that specific codehave been assessed.

Table 4.2 Enhanced ODC classification results

Defect Type	Qty	%	Defect Trigger	Qty	%	Defect Impact	Qty	%
Documentation	515	48.1%	Traceability/Compatibility	309	28.9%	Capability	308	28.8%
Function/Class/Object	203	19.0%	Test Coverage	227	21.2%	Maintenance	264	24.7%
Algorithm/Method	96	9.0%	Consistency/Completeness	206	19.3%	Reliability	252	23.6%
Checking	69	6.4%	Logic/Flow	119	11.1%	Documentation	157	14.7%
Interface	56	5.2%	Design Conformance	119	11.1%	Performance	39	3.6%
Build/Package/Environment	52	4.9%	Rare Situation	26	2.4%	Usability	28	2.6%
Assignment/Initialization	46	4.3%	Test Sequencing	16	1.5%	Requirements	9	0.8%
Timing/Serialization	33	3.1%	Standards Conformance	14	1.3%	Migration	8	0.7%
			HW/SW Configuration	13	1.2%	Standards	4	0.4%
			Recovery/Exception	10	0.9%	Installability	1	0.1%
			Other Triggers	11	1.0%			
Total	1070	100%	Total	1070	100%	Total	1070	100%

The '*Documentation*' defect type represents almost half of the defects and '*Function/Class/Object*' represents almost 20% of the defects. This can be justified by the fact that CSs highly depend on documentation and documented evidences to prove the accomplishment of requirements and standards and to ensure qualification/certification of the systems by external entities. '*Function/Class/Object*' identifies functionality implementation deficiencies, especially at implementation level.

'*Traceability/Compatibility*' is the most frequent trigger, although '*Test Coverage*' and '*Consistency/Completeness*' are quite frequent. This suggests that the most efficient defect triggers are the simplest and most logical ones, namely those related to traceability, reviews and testing activities. This is due to the nature of the artifacts under analysis that require extensive documentation and creation of evidences that are developed over lifecycle phases depending on the previous phases artifacts.

In terms of the impacts, four of them are very important, namely: '*Capability*', '*Maintenance*', '*Reliability*' and '*Documentation*' (in this order). It is normal that Capability (i.e. functionality) is the most affected property but, in such space CSs, maintenance has a significant importance as well as the reliability requirements (see Section 4.4.1 regarding the needs/objectives of the target systems).

4.4.4 Enhanced ODC Defect Impact Analysis

The ODC Impact analysis can be used to prioritize the defect types/triggers to identify the development and V&V activities that might conduct to the defects with a high impact in the system. As "high impact", we consider equally the impacts in Capability, Reliability, and Maintenance, as they are the most severe since they represent three essential requirements of critical space systems: functional quality, non-functional reliability assurance, and maintainability. Though, for the purpose of this work, we have considered the importance of impact as the frequency that the defects affect system capability, reliability, or maintenance.

The following graphs in this section represent the defect impacts as they have been originated by specific defect types, and also as they have been uncovered by specific defect triggers. The graphs provide an idea of the importance of defect types (related to root causes) and how defects that lead to specific impacts have been detected with specific triggers.

4.4.4.1 Type vs. Impact

Figure 4.3 shows the defect types that have a high impact in the system (affecting Capability, Reliability and Maintenance). Defects with impact in Capability (blue dashed line) are mainly related with Function/Class/Object, Documentation and Algorithm/Method types, confirming that the functionality specification/implementation, the documented artifacts and the design decision in what concerns algorithms and methods to apply are the main contributors to defects that influence the system capability.

Defects with impact in Reliability (orange dotted line) are originated from Documentation, Checking, Function/Class/Object and also Algorithm/Method defect types. In this case, there is a new defect type that contributes significantly to reliability issues: Checking. It is clear that reliability (including redundancy, fault detection/monitoring, isolation and recovery) is often implemented with checks and verifications and so the importance of avoiding this type of defects to guarantee higher reliability.

Defects with impact in Maintenance (gray line) originate essentially from Documentation defect type. This is an expected result due to the fact that maintenance depends on documented artifacts that include installation and download instructions, user and developer manuals, and maintenance procedures.

The prioritization related with the three impacts is presented in the Section 4.4.5, namely in Table 4.3.

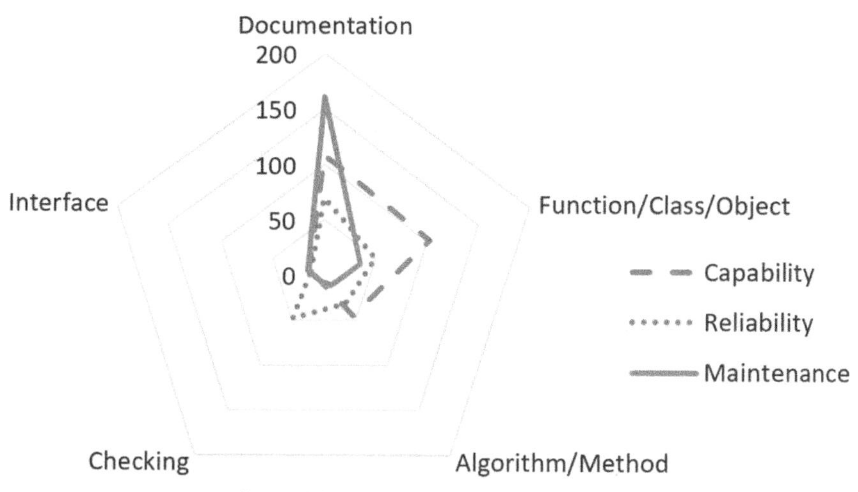

Figure 4.3 Defect type vs. defect impact.

Table 4.3 Summary of root causes for main defect types

Root Cause	Defect Types
Inefficient/insufficient reviews	Documentation; Function/Class/Object; Algorithm/Method; Checking; Interface
Ambiguous/missing/incorrect artifacts (documentation, requirements, design, tests)	Function/Class/Object; Algorithm/Method; Checking; Interface
Insufficient/Wrong tests (unit, integration, system, fault injection)	Function/Class/Object; Algorithm/Method; Checking; Interface
Limitations of the tools or toolsets that deal with documentation	Documentation
Lack of Completeness and consistency of system level (or previous phases) documentation	Documentation; Function/Class/Object; Algorithm/Method
Oversimplified documentation planning procedures	Documentation
Lack of time to produce, review and accept documentation artifacts	
Lack of importance given to some documentation artifacts	
Simplification of the product assurance processes related to documentation artifacts	
Limited engineers domain knowledge – lack of appropriate skills	Function/Class/Object; Algorithm/Method
Incomplete specifications in what concerns FDIR and erroneous situations	Checking
Lack of reliability and safety culture	Checking
Incomplete specifications in what concerns interfaces, environment and communications	Interface
Limited definition of the operation, usability, maintainability requirements	Interface
Lack of tools knowledge, programming languages, design languages	Function/Class/Object; Algorithm/Method
Version and configuration management procedures inappropriately implemented	Build/Package/Environment

4.4.4.2 Trigger vs. Impact

Figure 4.4 shows the defect triggers that allow detection of the defects with a high impact. The graph reinforces the importance of the 3 main triggers: a) Consistency/Completeness, b) Test Coverage, and c) Traceability/Compatibility as the most important (frequent) triggers (overall they allowed the detection of 77.0% of the issues). For this particular case, Reliability can

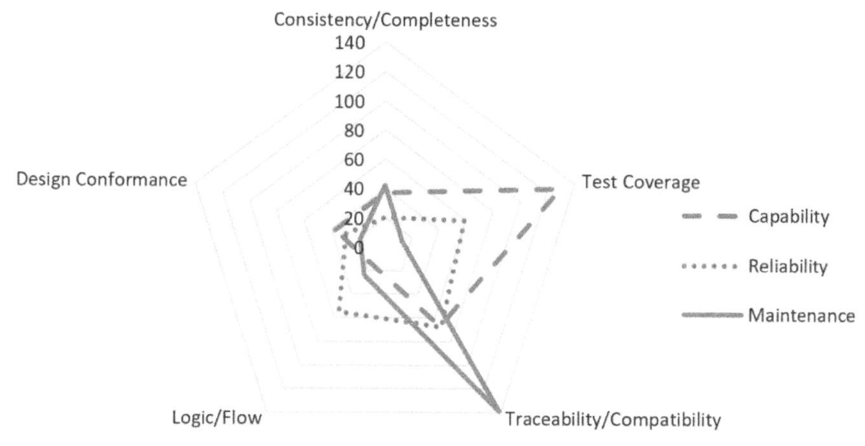

Figure 4.4 Defect trigger vs. defect impact.

be ensured with better Traceability/Compatibility analysis, Test Coverage and Logic/Flow analysis. Capability shall be assessed more efficiently with Test Coverage, Traceability/Compatibility assessment and Design Conformance Analysis. Maintenance defect impact can be mitigated with Traceability/Compatibility and Consistency/Completeness analysis.

The results of the prioritization related with these three impacts are presented in Section 4.4.5, Table 4.4.

4.4.5 Consolidation of the Root Cause Analysis and Proposed Improvements

The defects with impact on capability, reliability, and maintenance, identified in Section 4.4, represent 77% of the total dataset. From these, we considered the top 6 defect types and the top 5 defect triggers (Table 4.2) because they account for more than 90% of the defects with high impact. Then we were able to identify the main root causes for the most important defect types (Table 4.3) and the most important defect triggers (Table 4.4).

This analysis results on a list of the most important causes of the defects identified during ISVV, and for the most important causes of failure in the verification and validation activities during the development lifecycle. For high defects with impact, the listed causes show that software engineering processes, methods and tools require some adjustments in order to become more efficient to produce more dependable and safe systems. The identified root causes are all related to existing development and V&V activities that require more careful application, especially in what concerns schedule

Table 4.4 Summary of root causes for main defect triggers

Root Cause	Defect Trigger
Lack of traceability verification culture	Traceability/Compatibility
Lack or inefficient usage of tools that support traceability across lifecycle phases	
Lack of appropriate test planning and test strategy definition	Test Coverage
Lack or inefficient testing tool and testing environment support	
Incomplete tests specification and execution	
Review process related root causes	Document Consistency/ Completeness (Internal Document)
Documentation related root causes	Document Consistency/ Completeness (Internal Document)
Deficient usage of tools and applicable processes	Document Consistency/ Completeness (Internal Document)
Unclear or missing/incomplete specifications	Document Consistency/ Completeness (Internal Document); Logic/Flow
Ambiguous or unclear architecture definition	Logic/Flow
Lack of usage of tools that support data and control flow analysis	Logic/Flow
Inappropriate architecture support tools or tool usage	Design Conformance
Deficient specification or design artifacts	Design Conformance

and planning pressures (or we can call it strategies as well), rigor and caution on the application of engineering processes, and V&V activities importance. The quality/product assurance strategies and the guidance from applicable processes and required standards are essential to ensure that these root causes are minimized.

The root causes presented (in Tables 4.3 and 4.4) have been ordered according to expert knowledge and experience applicable to the high impact defects, and intend to provide a preliminary ordering in what concerns their contribution to the high defect impacts.

The identified root causes for defect triggers indicate that improvements to the current processes, both development (to avoid the introduction of defects) and V&V (to detect the defects within the phase they are introduced) might be possible. At a higher level, the leading safety standards might require additional guidance to support development and V&V in order to reinforce that the product/quality assurance (PA/QA), and safety and dependability assessments should be properly realized, reducing the amount of defects caught by ISVV. The proposed improvements are guidelines derived directly from the root causes summarized in Tables 4.3 and 4.4 and from domain and expert knowledge of the authors and industrial contributors to this work. Their intent is to fulfil the needs of the development and V&V processes in order to avoid the most important and more frequent defects as those in our dataset.

From the development perspective, based on Table 4.3, the following measures should be considered:

- Define/redefine appropriate review methods, processes and tools and enforce their application at every stage of the SDP;
- Implement automated documentation generation processes and tools to avoid inconsistencies between artifacts/lifecycle phases;
- Use tools that integrate and manage all the phases of the lifecycle, such as concept specifications, requirements, architecture, source code, tests, etc.;
- Introduce/use tools with automatic validations (documentation completeness, design consistency, code analysis, control and data flow analysis);
- Provide training to the engineering teams, to improve the domain knowledge, the system or interfacing systems knowledge, standards knowledge and techniques and tools practice;
- Promote workshops or meetings to present the specifications/requirements, to discuss and clarify them before advancing to the following phase;
- Introduce additional guidelines or even specific requirements (e.g., by defining and specifying the reasoning behind the standards requirements and how to achieve them in full conformance) in the applicable standards (PA/QA, version and configuration control and development).

From the V&V perspective, based on the results in Table 4.4, the following measures should be considered:

- Define appropriate test plans and strategies, especially unit and integration tests. The soundness of the test plans and strategies will reflect in the success of the validation;
- Ensure appropriate (or automated) traceability analysis at every stage of the development lifecycle;

- Improve the testing completeness, coverage and reviews;
- Implement non-functional tests (fault detection, fault injection, redundancy, etc.);
- Apply or develop tools to verify and validate the implementation and design compliance.

4.5 Conclusions

This chapter presented a defects assessment process based on a field study on root cause analysis of 1070 defects in space software projects.

We proposed a general procedure to derive improvement suggestions for the systems and the analysis process itself applying an improved ODC taxonomy and examining the defect types, triggers and impacts. We have also prioritized the root causes based on their importance by considering the impact of the defects on capability, reliability, and maintainability, and proposed generic solutions to implementation (to prevent defects) and V&V (to effectively detect defects) in order to avoid these defects in future projects.

The outcomes of the field study show that, although CSs are already guided by appropriate development and V&V techniques and processes, most of the defects are caused by an inefficient usage or implementation of these techniques and processes. Appropriate guidance, additional requirements and constraints, better test strategies and tools that are able to help in the application of the techniques and processes would be essential to obtain better results (less defects). ISVV was originally able to detect the 1070 defects but could still be enriched by applying the proposed V&V actions in order to avoid defects slippage.

References

[1] RTCA DO-178B. (1992). *(EUROCAE ED-12B), Software Considerations in Airborne Systems and Equipment Certification*. RTCA Inc., Washington, DC.

[2] Sai Global. (2011). CENELEC EN 50128: Railway applications – Communication, signalling and processing systems – Software for railway control and protection systems.

[3] ECSS. (2009). *ECSS-E-ST-40C, Space engineering – Software.*

[4] ECSS. (2009). *ECSS-Q-ST-80, Space Product Assurance – Software Product Assurance.*

[5] Silva, N., Lopes, R. (2012). "Overview of 10 Years of ISVV Findings in Safety-Critical Systems," in *2012 IEEE 23rd International Symposium on Software Reliability Engineering Work-shops (ISSREW)* (New York, NY: IEEE), 83.

[6] Silva, N., Lopes, R. (2012). "Independent Assessment of Safety-Critical Systems: We Bring Data!" in *IEEE 23rd International Symposium on Software Reliability Engineering Workshops (ISSREW)* (New York, NY: IEEE), 84.

[7] Silva, N., Lopes, R. (2012)."10 Years of ISVV: What's Next?" in *2012 IEEE 23rd International Symposium on Software Reliability Engineering Workshops (ISSREW)*(New York, NY: IEEE), 361–366.

[8] Silva, N., Lopes, R. (2011). "Independent Test Verification: What Metrics Have a Word to Say", in *1st International Workshop on Software Certification (WoSoCER)*, ISSRE, Hiroshima, Japan (New York, NY: IEEE).

[9] Silva, N., Lopes, R., Esper, A., Barbosa, R. (2013). "Results from an independent view on the validation of safety critical space system," in DASIA 2013, 14–16 May, Oporto, Portugal.

[10] EasterBrook, S. (1996). "The Role of Independent V&V in Upstream Software Development Processes", in *Proceedings of 2nd World Conference on Integrated Design and Process Technology (IDPT)*, Austin, Texas, December 1–4.

[11] Chillarege et al. (2013). *Orthogonal Defect Classification v 5.2 for Software Design and Code* IBM.

[12] Silva, N., and Vieira, M. (2014). "Towards Making Safety-Critical Systems Safer: Learning from Mistakes," in *ISSRE2014*, Naples, Italy.

[13] Silva, N., and Vieira, M. (2016). "Software for Embedded Systems: A Quality Assessment based on improved ODC taxonomy," in SAC 2016, Pisa, Italy.

[14] Copeland, L. "Software Defect Taxonomies". Available at: http://flylib.com/books/en/2.156.1.108/1/

[15] IEEE. (2010). *IEEE 1044-2009 Standard Classification for Software Anomalies*. Institute of Electrical and Electronics Engineers, New York, NY.

[16] Leszak, M., Perry, D. E., and Stoll, D. (2002). Classification and evaluation of defects in a project retrospective. *J. Syst. Softw.* 61, 173–187.

[17] Margarido, I. L., Faria, J. P., Vidal, R. M., Vieira M. (2011). "Classification of Defect Types in Requirements Specifications: Literature Review, Proposal and Assessment." 2011 6th Iberian Conference on Information

Systems and Technologies (CISTI) (New York, NY: IEEE), 1–6. Available at: http://ieeexplore.ieee.org/xpls/abs_all.jsp?arnumber=5974237

[18] ESA ISVV Guide, issue 2.0, 29/12/2008, European Space Agency.

[19] ISO/IEC 12207:2008 *Systems and software engineering – Software life cycle processes.*

[20] IEEE 1012-2004 – *IEEE Standard for Software Verification and Validation. IEEE Computer Society.*

[21] Jones, M. (2005). *Software Engineering: Are we getting better at it?* ESA Bulletin 121, 52–57.

[22] Leszak, M., Perry, D. E., Stoll, D. (2002). "A Case Study in Root Cause Defect Analysis," in *Proceedings of 22nd Intl Conf SW Eng (ICSE'OO)* (New York, NY: IEEE), IEEE CS Press, Los Alamitos, CA, 428–437.

[23] Lutz, R. (1993). "Analyzing Software Requirements Errors in Safety-Critical," in *Embedded Systems, Proc IEEE Intl Symp Req Eng* (New York, NY: IEEE CS Press), 126–133.

[24] Weiss, K. A., Leveson, N., Lundqvist, K., Farid, N., Stringfellow, M. (2001) "An Analysis of Causation in Aerospace Accidents," in Digital Avionics Systems, 2001. DASC. 20th Conference (New York, NY: IEEE).

[25] Seaman, C. B., Shull, F., Regardie, M., Elbert, D., Feldmann, R. L., Guo, Y., and Godfrey, S. (2008). "Defect Categorization: Making Use of a Decade of Widely Varying Historical Data," in *Proceedings of the Second ACM-IEEE International Symposium on Empirical Software Engineering and Measurement* (New York, NY: ACM), 149–57. Available at: http://dl.acm.org/citation.cfm?id=1414030

[26] Neufelder, A. M. (2012). *The top ten things that have been proven to impact software reliability.* Available at: http://www.softrel.com/downloads/TopTen.pdf

[27] Rao, R. (2014). *Root Cause Defect Classification (RCDC) for Documentation Defects.* Available at: http://www.stc-india.org/conferences/2014/presentations/Root%20Cause%20and%20Defect%20Classification%20for%20Documentation%20Bugs%20-%20Ramaa%20Rao.pdf

[28] Kumaresh, S. and Baskaran, R. (2010). Defect Analysis and Prevention for Software Process Quality Improvement. *Int J. Comput. Appl.* (0975–8887), 8.

[29] Silva, N., Vieira, M., Ricci, D., Cotroneo, D. (2015). "Assessment of Defect Type influence in Complex and Integrated Space Systems: Analysis Based on ODC and ISVV Issues," in *2015 IEEE International Conference on Dependable Systems and Networks Workshops (DSN-W)* (New York, NY: IEEE), 63–68.

5

Framework for Automation of Hazard Log Management on Large Critical Projects

Lorenzo Vinerbi[1] and Arun Babu Puthuparambil[2]

[1]Resiltech s.r.l., Pontedera (PI), Italy
[2]Robert Bosch Center for Cyber Physical Systems, Indian Institute of Science, Bangalore, India

5.1 Introduction

A hazard (HZ) is any situation that could cause harm to the system or lives. HZ depends on the system and its environment, and the probability of the HZ to cause harm is known as risk. HZs are analyzed by identifying their causes and the possible negative consequences that might ensue. For example, the dangerous failure of a traffic signal could be caused by a logic error in the traffic signaling controller's software program. The consequence could be conflicting traffic flows simultaneously receiving green signals.

A hazard log (HL) is a database of all risk management activities in a project. Maintaining its correctness and consistency on large safety/mission critical projects involving multiple vendors, suppliers, and partners is critical and challenging. IBM DOORS [1, 2] is one of the popular tool used for HZ management in mission critical applications. However, not all stake-holders are familiar with it. Also, it may not always feasible for all stake-holders to provide correct, well structured, and consistent HZ data. IBM DOORS have been reported to be useful in managing DO-178 compliance for avionics [3]. Also, HL in DOORS allows capabilities for tracing requirements and test results. However, DOORS has steeper learning curve and is difficult to use by common people and beginners [4]. Also, they lack validation capabilities [5]. Custom checks may require difficult to use plug-ins which are not generic. This complexity makes it difficult to maintain the rules; preventing reuse in other projects.

This chapter demonstrates a modular and extensible way to specify rules for checks locally at the stake-holder side, as well as while combining data from various parties to form the HL. The HZ-LOG automatization tool simplifies the process of HZ data collection on large projects to form the HL, while ensuring data consistency and correctness. The data provided by all parties are collected using a template containing scripts. The scripts check for mistakes/errors based on internal standards of company in charge of the HZ management. The collected data is then subjected to merging in DOORS, which also contain scripts to check and import data to form the HL.

The requirements of HL tool are:

(i) Perform checks of incoming data from vendors and partners;
(ii) It shall allow to collect and keep log for all information related to identified HZs (and related identified mitigations), structuring information accordingly;
(iii) It shall be possible to manage the status of the HZs and related mitigations, allowing for the control of risk. Only allowed HZ status transitions shall be possible and logging of the related status transition activity shall be kept in the tool for traceability purposes;
(iv) Only RAMS specialist are allowed to manage HZs being necessary no different user profiles for the management of HZs in the tool;
(v) A function of the tool shall allow to extract the "current" status of the project system HL by allowing the creation of documentary reports containing the set of necessary information about the predicted HZs, mitigations identified, and the status of all related risk control activities.

5.1.1 Brief Introduction on DOORS

IBM Rational DOORS is an enterprise-wide requirements management tool, designed to link and manage diverse textual and graphical information to ensure a project's compliance to specified requirements and standards. It represents a layer to perform:

(i) Import documentation into a DB in order to convert free text into requirements;
(ii) Maintain such requirements during the time;
(iii) Relate requirements belonging to different documents (or level of detail);
(iv) Relate requirements to other artefact (e.g., test specification or report).

Due to its features, it is widely adopted in different domain as reference tool to manage requirements and HL.

5.2 Approach

All the activities described in the previous sections lead to a set of HZs and mitigations; which in the end allow to guarantee the safety all along the lifecycle (see Figure 5.1).

The mitigations identified in PHA [6] and SHA [7] shall be evaluated, along with design changes, on a continuing basis, to ensure that risk associated to HZs has been eliminated or lowered to an acceptable or practicable level. The result of this activity shall be stored in the Hazard Log Tool. Some other activities may provide results to be logged, e.g. design implementation schemes, design analyses, test specifications and test reports etc. Whilst main HZ analyses are planned by the project's safety plan, [7] other safety analyses and project activities providing results to be logged in the HL have no plan. It is the Safety Organization's responsibility to log the outcome of safety activities when resulting new HZs as well as to record all the information necessary to provide final evidence of safety.

A template with configuration and script is created and sent to all participants in the project. Template fields are listed and explained in Table 5.1. Table 5.2 reports possible configurations for mapping DOORS fields into excel ones, while Table 5.3 is an example of configuration for excel template that specifies allowed combination of hazard frequency, severity, and risk

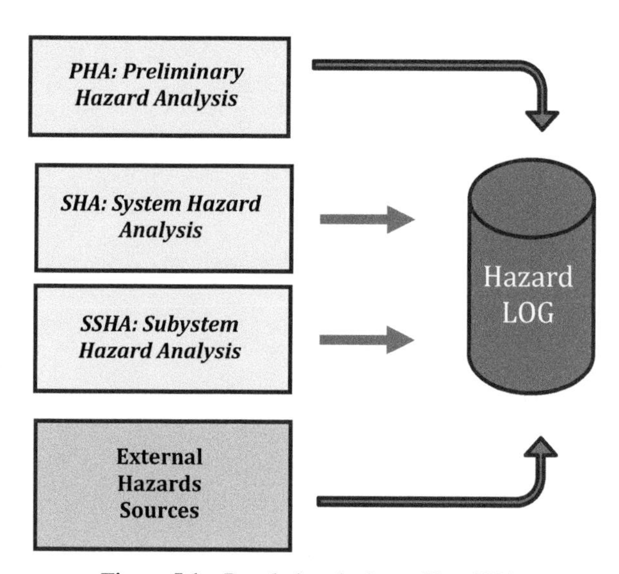

Figure 5.1 Populating the hazard log (HL).

Table 5.1 Hazard analysis template

HLS ID	Field Name	Format	Description
1	Hazard ID	A specific format has to be defined	It is the unique identifier for a hazard.
2	Hazard Opening Date	A specific format has to be defined	This field contains the date in which the hazard has been opened.
3	Hazard Closure Date	A specific format has to be defined	This field will contain the date in which the hazard closes.
4	Hazard Source	Text	Initial generic source from which the hazard was identified.
5	Hazard Description	Text	A complete exhaustive description of the hazard.
6	Hazard Cause	Text	All possible failure modes of functions/subsystems/ equipment/components which could lead to the hazard.
7	Hazard Consequence	Text	It is the possible accidents to which the hazard could lead.
8	Hazard Event	Usually each hazard is categorized following a limited list of possible hazard event, in order to ease maintenance and analysis of the results	This field reports the top level event (or a combination of events) resulting from the hazard.
9	Hazard Initial Frequency	One out of # possible values (they depends on the project, e.g., "Incredible", "Improbable", "Remote", "Occasional", "Probable," or "Frequent")	This field evaluates the initial frequency of the hazard, based on previous experiences, previous evaluations, expert judgment, statistical analysis and by considering the existing mitigations of legacy system and so it will be based on the data/information already available.
10	Hazard Initial Severity Level	One out of # possible values (they depends on the project, e.g., "Catastrophic", "Critical", "Marginal" or "Insignificant")	This field evaluates the severity of the consequences related to the hazard, based on previous experiences, previous evaluations, expert judgment, statistical analysis and by considering the existing mitigations of legacy system and so it will be based on the data/information already available.

11	Hazard Initial Risk Valuation	One out of # possible values (they depends on the project, e.g., “Undesirable”, “Intolerable”, “Tolerable” or “Negligible”)	It is the combination of initial consequence and initial frequency. It establishes the level of risk generated by the hazardous event.
12	Hazard Final Frequency	One out of # possible values (they depends on the project. e.g., “Incredible”, “Improbable”, “Remote”, “Occasional”, “Probable” or “Frequent”)	In this field we report the final residual frequency of the hazard.
13	Hazard Final Severity Level	One out of # possible values (they depends on the project. e.g., “Catastrophic”, “Critical”, “Marginal” or “Insignificant”.)	This field evaluates the final residual severity of the consequences related to the hazard.
14	Hazard Final Risk Evaluation	One out of # possible values (they depends on the project. e.g., “Undesirable”, “Intolerable”, “Tolerable” or “Negligible”)	It is the final combination of residual consequence and frequency.
15	Hazard Status	One out of four possible values: “Open”, “Solved”, “Deleted” or “Closed”.	It is the status of hazard.

Table 5.2 An example configuration of hazard log tool ("Hazard Log Field" are the fields in DOORS, "HA" is the fields in Excel, and "Type" indicates where the field can be found (HZ, 'hazard'; MT, 'mitigation'; BH, 'can be found in both')

Hazard Log (HL) Field	Type	HA	Id
Hazard Log Id	BH	Hazard Log Id	1
Hazard Opening Date	HZ	Hazard Opening Date	2
Hazard Revision Id	HZ	Hazard Revision Id	3
Hazard Closure date	HZ	Hazard Closure date	4
Hazard Consequence	HZ	Hazard Consequence	5
Hazard Frequency Pre Mitigation	HZ	Hazard Frequency Pre Mitigation	6
Hazard Status	HZ	Hazard Status	7
Mitigation Id	BH	Mitigation Id	8
Mitigation status	MT	Mitigation status	9

Table 5.3 Example configuration for Excel scripts

Hazard Frequency Pre Mitigation	Hazard Severity Level Pre Mitigation	Hazard Risk Evaluation Pre Mitigation
	Allowed Words	
F0-Frequent	S4-Disastrous	Intolerable
F1-Probable	S3-Catastrophic	Undesirable
F2-Occational	S2-Critical	Tolerable
F3-Remote	S1-Marginal	Negligible

evaluation tool. This template is designed in MS-Excel, which allows running of scripts/macros. These macros are written considering requirements of the project.

The database consists of a collection of HZ records (one record for each identified HZ) and a collection of the mitigation action records related to the identified HZs. Each HZ record contains the information regarding the HZ such as: Hazard identification, Hazard Revision Number, Identified in phase, Hazard originator's code, Operating mode, Hazard description, etc., as per the company and project specific standard. Also, the systems and subsystems have to identify all necessary mitigations to the identified HZs so the associated risk is eliminated or ALARP (as low as reasonably practicable) according to the risk categories definitions and as explained in the safety cases. For each HZ, mitigation actions are specified to control the risk to ALARP. Each mitigation record contains information such as: Mitigation ID, Mitigation Revision, Mitigation Revision date, Mitigation Description, Applied to phase, Mitigation Status, etc. Since, each project has different needs, check of data consistency and correctness rules are needed to generate correct HL. Hence, a template and set of rules are created in MS-Excel. The rules are based on high-level requirements of standards of company in charge of HZ management, written in the form of scripts [8]. Each participant receives the template, and it is filled out with HZ data and it is thoroughly checked with Excel scripts (Figure 5.2, Figure 5.3, Figure 5.4, Figure 5.5). Once all checks are passed, it is compliant with the company and project standards. It is then sent to a central place to merge and form the *HL*. The merging of data from Excel format to DOORS is done through custom scripts which validates the data columns for correctness and consistency (Figure 5.6). Each HZ data from a participant is checked for consistency using scripts in DOORS and are integrated to form HL if no errors are found. Often Excel file consist of more fields than that of DOORS, they are either discarded or used for computation. A second script checks if a previous version of the file was uploaded yet, in such case HL is updated. Finally, the HZ log sheet is produced containing: Hazard identification, Hazard revision number, Hazard originator's Code, Hazard description, Hazard Owner, Party to act, Hazard Comments, Mitigation Comments, etc. Several fields are marked as NULL; as they will be entered during the lifetime of the system.

The cost-effectiveness of the HL management process has been achieved by the following scripts:

(i) Scripts to be used jointly with MS Office tool suite in order to make simple checks, and to reduce the number of errors introduced into the DOORS DB;
(ii) Scripts to be used in DOORS in order to ease import from excel file, update and export. Concerning the support for MS Office, the scripts were created implementing the following checks of interest for an HL:

- Concerning the hazards:
    - each hazard shall have a unique identifier;
    - each hazard shall have a non-empty "consequences", "causes", and "status";
    - each not cancelled hazard shall have a risk evaluation pre-mitigation;
    - each not cancelled hazard having a risk level pre-mitigation higher than tolerable shall have a risk evaluation post-mitigation;
    - when risk evaluation is applied the risk matrix shall be respected;
    - each hazard having status different from "cancelled" or "open" shall have a mitigation.

- Concerning the mitigations:
  - each mitigation shall have a unique identifier;
  - each mitigation shall have a non-empty “description”, “assigned to”, “status”.
- Concerning the traceability:
  - if “Mitigation Implementation (reference)” field is not empty, check trace on document list;
  - if “SRAC” field is not empty, check trace on SRAC list;
  - if “RTM” field is not empty, check trace on RTM list;
  - in case of structured HL (i.e. HZ and mitigation separated tables) – coherence checks like:
    - Does the mitigation referred in HZ table have at least an existing HZ?
    - Does all the mitigations referred in HZ table exists in the mitigation table?

5.3 Case Study

The proposed approach has been applied to four different critical projects where each project has 6–10 suppliers, and each supplier produced HZ analysis with 200–400 rows and the merged HL of ~2000 rows for each project.

In order to evaluate the correctness and the improvement given by the scripts, we used them in different real project in order to appreciate how it is used by different teams working on different contexts. In particular we used four projects related to the Railway domain, concerning metro lines to be installed in different cities.

The main characteristic of the different project are shown in the below table.

Metro Line	Team Size	No. of Involved Subsystems	Project Duration	No. of Hazards Composing the HL
Metro X01	3	9	2015–2016	~1600
Metro X02	4	10	2015–on going	~2000
Metro X03	3	8	2015–on going	~1400
Metro X04	5	10	2014–on going	~2500

Scripts have been used during the different phases of the safety lifecycle. The scripts related to MS Office have used in the early stages to evaluate first drafts (/releases) of the files coming from suppliers. The feedbacks from the different teams are quite similar:

(i) No. of syntactic errors contained in the files given by the suppliers are drastically reduced (90%);
(ii) Time spent in reviewing (just from syntactic point of view) is drastically reduced (70%);
(iii) This goal has not been reached in a single step, indeed most of the suppliers complained on the low usability of the scripts. This is something we expected indeed they are just first releases to have feedback "from the field", so the user interface was not good enough to be reasonably used without some initial difficulties.

Scripts related to MS Office have been then used to verify correctness of the integrated HL (the one composed joining the different HAs from suppliers). Scripts related to DOORS have been used to import system HL in DOORS. In this case, feedbacks from the different teams differs. Most of the teams noticed a real improvement in using such scripts, since they:

(i) reduce time related to import activities;
(ii) make people, who are not familiar with DOORS, capable to easily use it; and
(iii) reported a real decrease in time connected with import activities (up to 80%).
(iv) One team reported no real improvement in using such scripts. This is because:
    (a) people present in the team are very skilled on DOORS (they already have their own processes to easily import HAs on it);
    (b) the presence of template for HA are really hard to be managed. This led a lot of error related to configuration of the script, which took more time to be solved.

Scripts related to DOORS have also been used to update HL. In this case, feedbacks from the teams were not so good. Indeed, most of them reported difficulties in applying the process to be used in order to correctly keep track of the different change in HL. This has led us to re-consider this phase from the scratch and changing such approach in the future projects.

5.4 Conclusion

As HL is the database for all HZ/risk information and is updated throughout the project life-cycle, it is critical that the HZ analysis data is correctly and consistently merged. Especially, in large projects having multiple partners/vendors. In the current study, the proposed approach has been found to be useful in reducing mistakes in HZ analysis. Also, it has been found to reduce the amount taken to create the HL. The use of automatic checks paves the way for correct tracking of risk and HZ analysis activities for large critical projects. More specifically:

(i) All the excel sheets from all participants have been automatically imported into the DOORS tool;
(ii) It has been observed that a significant reduction in the number of non-conformities presents in the document provided by the different suppliers.
(iii) The time required to merge data to form HL is reduced by ~30%.
(iv) Engineers in the main company are now more likely to use DOORS, since the offered framework, allowing them to easily interact with it. This also resulted in increase in quality of the project. The proposed approach has been found to be generic and suitable to all critical systems.

5.5 Tool Screenshots

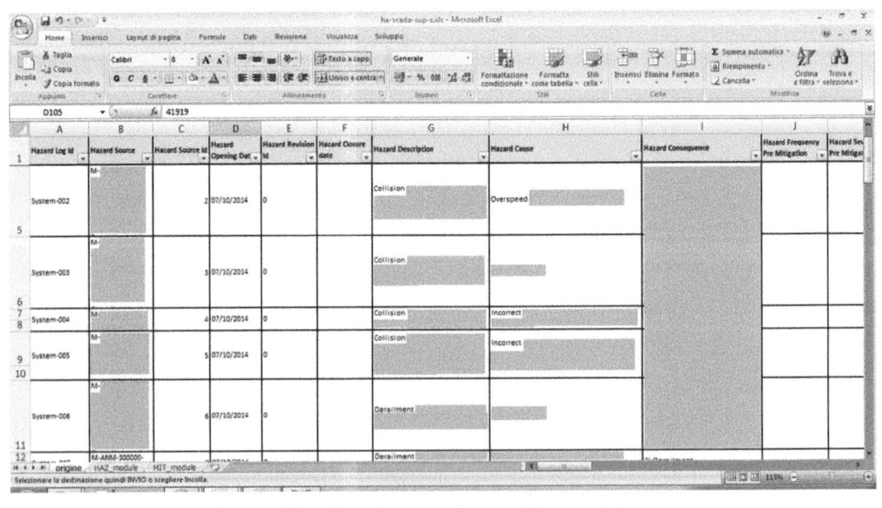

Figure 5.2 Excel sheet of one of the participants.

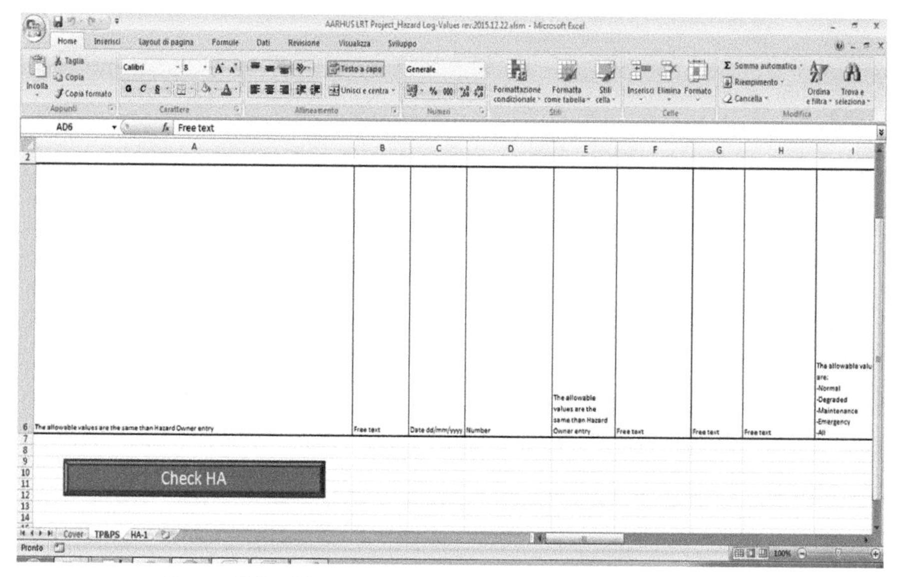

Figure 5.3 Checking of HA data through MS Excel scripts.

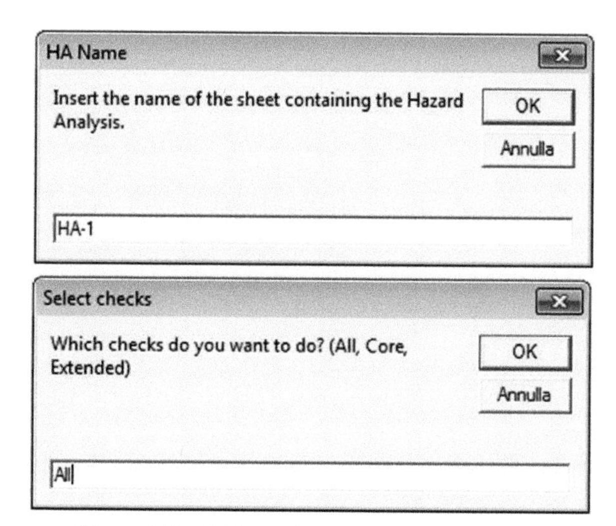

Figure 5.4 Dialogue boxes of MS Excel scripts.

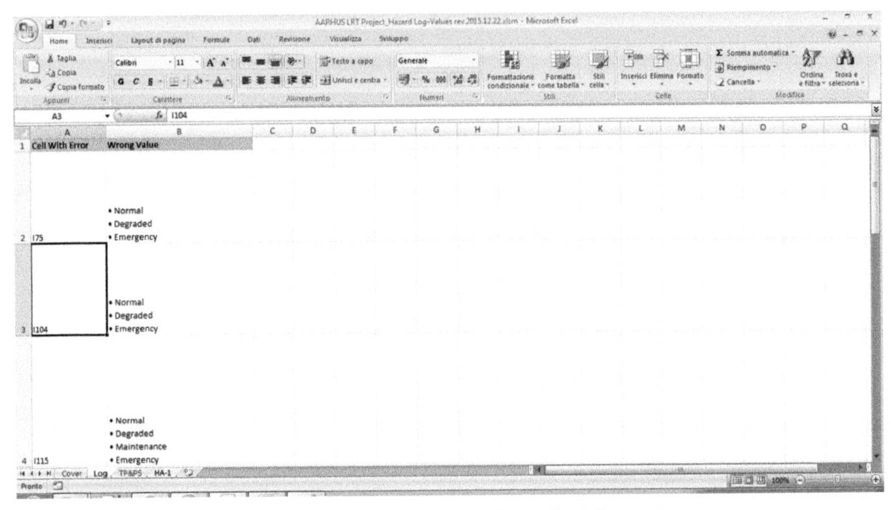

Figure 5.5 Errors caught in HZ analysis by scripts.

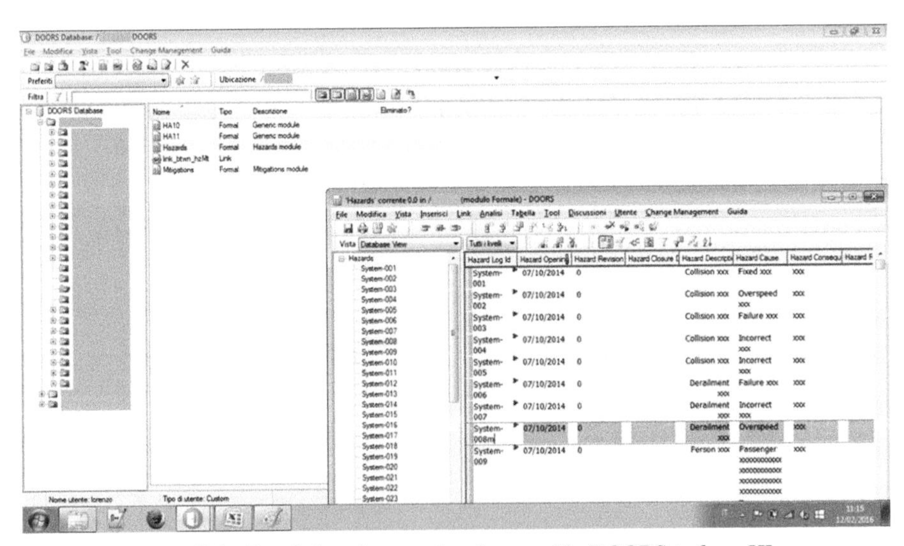

Figure 5.6 Excel sheet imported and merged in DOORS to form HL.

References

[1] IBM. (s.d.). (*Rational DOORS Family*). Available at: http://www-03.ibm.com/software/products/en/ratidoorfami (accessed on 15 February 2016).

[2] Dave, H., and Saeed, B. (2009). "Hazard Management with DOORS: Rail Infrastructure Projects," in *Safety-Critical Systems: Problems, Process and Practice* (London: Springer), 71–93.

[3] Çakmak, K. M. (2013). Managing DO-178 Compliance with IBM Rational Platform. *J. KONBiN*, 25, 59–74.

[4] Lööf, R., and Pussinen, K. (2014). *Visualisation of requirements and their relations in embedded systems*. Uppsala University. Sweden.

[5] Dibbern, J., Geisser, M., Hildenbrand, T., and Heinzl, T. (2009). Design, implementation, and evaluation of an ICT-supported collaboration methodology for distributed requirements determination. Working paper.

[6] Pasquale, T., Rosaria, E., Pietro, M., Antonio, O., and Segnalamento Ferroviario, A. (2003). "Hazard analysis of complex distributed railway systems," in *Proceedings of 22nd International Symposium on Reliable Distributed Systems, 2003*, 283–292. doi: 10.1109/RELDIS.2003.1238078

[7] Chapra, S. (2003). *Power Programming with VBA/Excel*. Upper Saddle River, NJ: Prentice Hall.

[8] Gowen, L. D., Collofello, J. S., and Calliss, F. W. (1992). "Preliminary hazard analysis for safety-critical software systems," in *Eleventh Annual International Phoenix Conference on Computers and Communication [1992 Conference Proceedings]*, Scottsdale, AZ, USA, 1992, 501–508. doi: 10.1109/PCCC.1992.200597

6

Cost Estimation for Independent Systems Verification and Validation

András Pataricza[1], László Gönczy[1], Francesco Brancati[2], Francisco Moreira[3], Nuno Silva[3], Rosaria Esposito[2], Andrea Bondavalli[4,5] and Alexandre Esper[3]

[1]Dept. of Measurement and Information Systems, Budapest University of Technology and Economics, Budapest, Hungary
[2]Resiltech s.r.l., Pontedera (PI), Italy
[3]CRITICAL Software S.A., Coimbra, Portugal
[4]Department of Mathematics and Informatics, University of Florence, Florence, Italy
[5]CINI-Consorzio Interuniversitario Nazionale per l'Informatica-University of Florence, Florence, Italy

Validation, verification, and especially certification are skill and effort demanding activities. Typically, specialized small and medium enterprises perform the independent assessment of safety critical applications. Prediction of the work needed to accomplish them is crucial for the management of such projects, which is by nature heavily dependent on the implementation of the V&V process and its support. Process management widely uses cost estimators in planning of software development projects for resource allocation. Cost estimators use the scoring of a set of cost influencing factors, as input. They use extrapolation functions calibrated previously on measures extracted from a set of representative historical project records. These predictors do not provide reliable measures for the separate phases of verification, validation and certification in safety critical projects. The current chapter summarizes the main use cases and results of a project focusing on these particular phases.

6.1 Introduction

Verification and validation (V&V), and Independent Software and System Verification and Validation (ISVV) are crucial in critical system development. However, the execution of such activities underlies strict time and budget limits and depends on input elements delivered by the entity performing the design and implementation of surprisingly changing quality.

Our research (described in details in CECRIS [1, 2]) focused on synthetizing a method for estimating the cost and focusing the effort of V&V activities in order to make these critical steps more managed and foreseeable in terms of time and cost aspects.

This necessitated an appropriate, organized, and scientifically well-founded working methodology.

The traditional software and system design industry has well-elaborated methods for the assessment of the aspects of efficiency and quality w.r.t. the human actors, technology used and project management background of the companies. Cost estimators (CEs) use the scoring of a set of cost influencing factors, as input. Their respective predictor uses extrapolation functions calibrated previously on measures extracted from a set of representative historical project records.

6.1.1 ISVV Workflow

The domain standards [3–8] define the project lifecycle and forms part of the design workflow and its phases. They envisage a purely top-down process starting with the requirements for implementation and checks. The individual phases are self-contained in the terms of design and V&V. Figure 6.1 presents a schematic view on the interoperation of the development and V&V phases.

However, the practice indicates, that in many cases there are deviances of this idealistic model. For instance, changes in the specification triggered by the end user during the design may result in iterations.

Constraints related to delivery time force occasionally corrections executed in parallel in phases intended to be sequential by the idealistic process model. We suppose a typical "waterfall" development, which, however, has its disadvantage that decisions taken in the project are often not

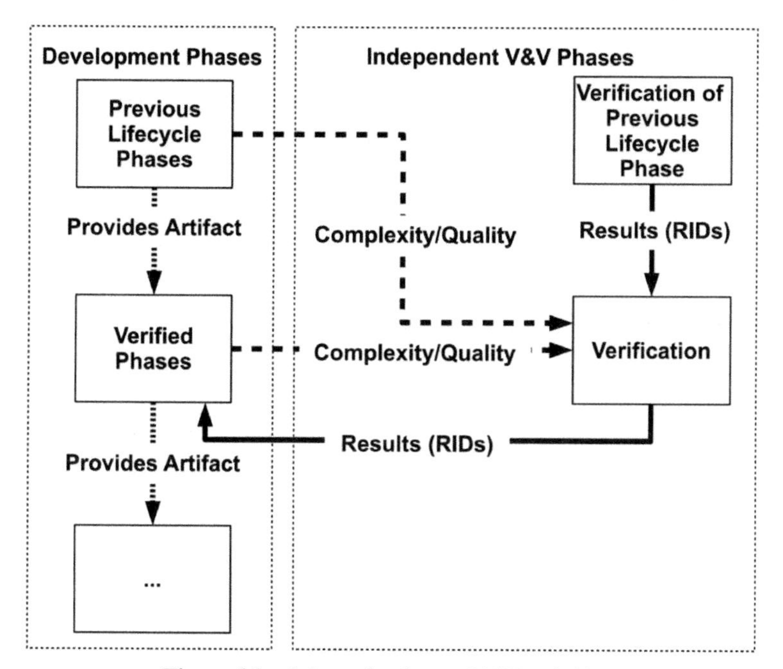

Figure 6.1 Schematic view on V&V activities.

reflected by artifacts (e.g., design documentation) created in an earlier project phase.

Even in critical projects, due to resource constraints, early deliverables are often not updated which may result in inconsistent documentation.

Note that the Figure 6.1 shows a simplified partial view of the process presented in Pataricza et al. [9].

Verification and validation activities of a particular phase and design activities in the successor phase frequently show a similar temporal overlap, for instance, coding starts before the completion of design verification.

If design verification detects faults later, feedforward change management has to take care to correct the design and update the specification of the already ongoing coding activity.

Furthermore, latent faults escaping the V&V of their respective earlier phase trigger feedback loops and multi-phase correction actions affecting preceding project deliverables as well.

Review Item Discrepancy (RID) is the output measure of a V&V activity. A RID is an issue, identified by a reviewer who is uncompliant with a requirement, a review objective or a design goal.

Our suggestion is to facilitate effort estimation and "spot out" problematic parts of the target artifact of verification by applying complexity/quality metrics. These metrics can be retrieved right at the beginning of verification; also, such metrics for previous project deliverables (e.g., code metrics in the case of test verification) are reusable.

6.1.2 Objectives

Cost estimators are fundamental elements all along the lifecycle of a project from the proposal definition through the project execution and *a posteriori* evaluation. Their main purpose is assuring a timely execution of the target project without overspending. A variety of general-purpose CE methodologies exists supporting the project management activities by predicting Key Performance Indicators (KPI), such as time/effort/cost/quality/risk.

Similarly, different measures assure the compliance of an ISVV or certification process with the standards. However, the objective of an assessor is an effective V&V and certification process among the standard-compliant ones. Effectivity means here both productivity and quality.

Rough estimates, and rules of thumbs like approximating testing related costs as by around 40% of the overall development project costs, as typical for non-critical applications do not deliver reliable results for safety-critical ones.

Academic experts performed a thorough going evaluation of the current practice of the ISVV companies performed for third parties. They analyzed numerous historical project logs with a particular focus on the cost aspects to identify the main factors influencing the efficiency of the projects and the dominant quality and productivity bottlenecks. One main observation was that automatic quality checking of incoming project artifacts has a high priority, as this is the core factor determining the efficiency of the entire assessment projects.

The primary objective of the research based on the results of the analysis was checking the validity of the original approach of creating general-purpose CEs for the purpose of ISVV-related effort prediction.

The most important use case is the improvement of the ISVV process. Accordingly, the focus of the evaluation was on the "what-if" – like analysis of the impacts of factor-by-factor changes in the process factors. This way, sensitivity analysis can predict the relative improvement expected, for instance, by the introduction of a new technology.

6.1.3 Approach

The enrichment of the ISVV workflow design by effort metrics necessitates a proper cost predictor. The literature refers to a large number of general-purpose and dedicated approaches.

Flexibility, understandability and possible accuracy were the main selection criteria in designing a CE specific for V&V.

The COCOMO family of CEs served as the starting point, as this is popular and relies on an open model. Note that the different versions of the CEs covering different use cases take nearly twenty factors as input to generate a cost estimate (as detailed in the next section).

The moderate number of historical project log does not even adequately sample this vast parameter space allowing the generation of an entirely new CE dedicated to ISVV.

Accordingly, our approach consisted of the following main steps:

- estimation of those general factors related to software development processes which are independent of the peculiarities of ISVV;
- cross-validation of the measured predicted by the general-purpose CE and the logged efforts in the ISVV sample set;
- checking the set of input factors for completeness and potential revision of the definition according to ISVV.

6.2 Construction of the ISVV Specific Cost Estimator

Software project management offers a huge variety of approaches and tools for cost estimation. As no single one is detailed enough to describe the specific V&V processes related to critical systems, we started the elaboration of an ISVV-specific CE model referred further as CECRISMO [10].

The COCOMO family of CEs served [11–13] as the starting point of the designated CECRISMO. The COCOMO family of CE has a widespread industrial acceptance and use including (embedded) system design. Members of it support different project types and process models including mixed hardware–software development, component reuse, and integration-based system design, etc. All COCOMO styled predictors rely on an open model. Thus, the adaptation can follow the usual process of the creation of a new member of this family. Most members of the family have an open source tool support.

6.2.1 Structure of the Cost Predictor

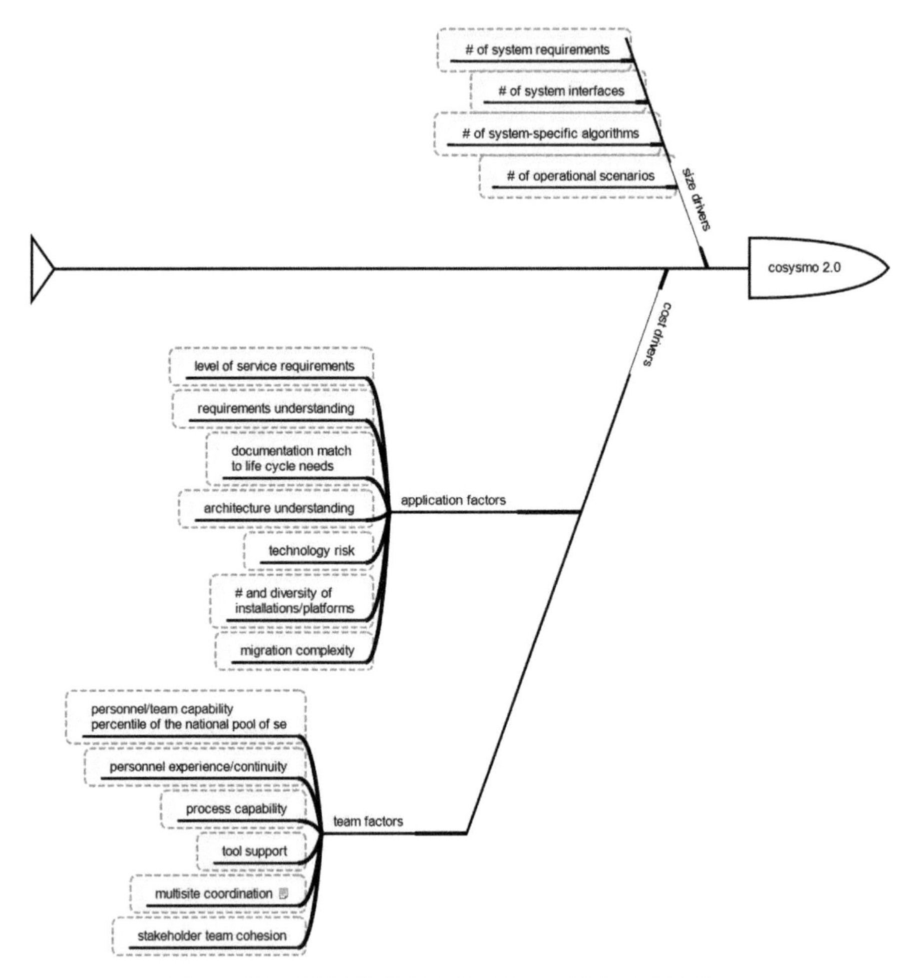

Figure 6.2 COSYSMO 2.0: Size Drivers/Effort Multipliers.

All the members of the COCOMO family (incl. COSYSMO) share a similar formula for cost prediction (see Figure 6.2).

$$\text{Person Months}_{\text{Nominal Schedule}} = A(\text{cost driver})^{E} \prod \text{effort multiplier},$$

where

- $Person\ Months_{Nominal\ Schedule}$ is the estimated effort in man-months needed for the end-to-end execution of the project (nominal schedule),

- "*size driver*" is an estimated metrics (a single real number) of the size and complexity of the target of the project ("*What is the output of the development project?*").
- For instance, early COCOMO cost predictors used the estimated number of source line of codes or function points as size estimators in pure coding projects. As not all elements within a particular category result in the same amount of efforts, difficulty categories express the differences in efforts and the total numbers are calculated as weighted sums. These weights are in the range of one order of magnitude for each step of a difficulty category. Similarly, the decreased efforts due to reuse of already existing artifacts is weighted by a reduction factor.
- "*effort multiplier*" expresses the efficiency of the development process from technology, skills, development infrastructure and organizational aspects ("*How effective is the output of the development project elaborated?*");
- The evaluator expert assigns to each individual effort influencing factor a grade (an ordinal measure) **Very low/Low/Nominal/High/Very high,** which in turn is converted with an aspect dependent empirical constant vector into a relative effort multiplier (for instance an effort multiplier metric less than 1 corresponds to a speedup w.r.t. to the nominal case).
- A and E are calibration constants estimated during curve fitting to the data in the calibration set.

6.2.2 Cost Drivers

COSYSMO targeted the entire system development process in the large. When designing a new member of the COCOMO family, the standard procedure is an analysis and if necessary of the original model simultaneously keeping the core of its original factors and prediction methodology.

The following evaluation targets the creation of a CE corresponding to ISVV in focus. This analysis addresses the following central questions:

- How appropriate are the input factors defined originally for the end-to-end process for ISVV (completeness and scoring methodology)?
- Is there a necessity for the re-interpretation of the factors originally defined for end-to-end development to the peculiarities of ISVV?

6.2.3 Focal Problems in Predicting Costs for ISVV

The primary reason of the non-existence of ISVV specific CEs is the lack of a sufficient number of available project logs and other processable artifacts supporting statistically justified generalized statements.

In addition, the efforts needed for ISVV heavily depend on the quality of the input artefacts supplied by the separate design entity. This way, the input quality has an at least equally important influence, like the size and complexity of the system under assessment.

These important interdependencies necessitate a re-interpretation of the individual factors.

General-purpose CEs have to cover a broad spectrum of human skill levels ranging from a post-secondary level programmer to domain experts. Accordingly, they apply a rough-granular scoring range covering this wide set. ISVV, a highly demanding task is typically carried out by SMEs having a well-educated staff frequently having a higher academy degree and a long industrial expertise. This way, the staff in ISVV belongs to the top few scores related to human factors in general-purpose CEs, and the expressive power of these factors becomes insufficient due to the low resolution.

6.2.4 Factor Reusability for ISVV-Related CE

The **size drivers** express the scale of the system under development adequately, but the size alone is insufficient to determine or approximate the efforts necessary for ISVV.

In the case of system requirements and artifacts corresponding to them, the impact of safety criticality on ISVV-related efforts needs refined weighting methods and scores. Critical requirements, components, etc., necessitate a thoroughgoing analysis as defined in the standards.

Moreover, standards define variation points in the checking process. Accordingly, here the weight of the factor needs a fitting to the actual process instance chosen for the actual object of ISVV instead of a global weight used in the original CE.

Similarly, as ISVV is a follow-up activity of a design phase the design quality heavily influences the amount of work to its execution.

The quality of the outputs of some V&V preliminary activities (e.g., design the V&V Plan, Requirement Verification, and Hazard Analysis) are major effort drivers for rest of the V&V cycle. Modeling this behavior will also allow to perform ROI analyses at early stage of the V&V process and/or continuously monitoring the cost-quality tradeoff of the overall V&V cycle.

The notion of quality has a double meaning here. On the one hand, it refers to the care of the preparation of the ISVV input artifacts (how well is a design document structured and formulated). The quality of the input artifacts has a direct impact on the problem size (how many items needs the review report to detail).

On the other hand, it covers technical aspects (like the testability of code). The expressive power of the grading system is insufficient, and this issue is subject to a detailed analysis in a subsequent section.

6.2.5 Human and Organizational Factors

An important obstacle originates in the difference regarding the organizational background.

Companies developing non-critical applications typically follow an end-to-end in-house approach. Mono-company development assures independence at the team-level within the company boundaries, if required by the standards, at all. Both the development and V&V teams share the same enterprise culture regarding skills, organizational and technology aspects in case an entirely in-house process. This way the assumption of having a nearly homogenous culture and quality along the entire process is highly probable. Experts performing the scoring of the individual organizational and technology-related cost factors can typically do it at the level of the company.

On the contrary, ISVV relies on two organizationally independent, but to a given extent collaborative partners each having his separate culture. Moreover, ISVV performed for different clients faces different cultures at the side of the designer companies.

The group of team factors has nearly the same semantics as in the original model.

However, ISVV is typically a much more complex activity using different methods and tools as a traditional development process. This way, for instance, human and technical factors are related to the individual activities instead of using flat, global grading. If an ISVV process consists of a mix of peer review and formal analysis, tool support, personal experience, etc. all have to be evaluated at the resolution of these individual activities instead of a single approach based scoring. Mostly SMEs having a limited number of experts perform ISVV. Occasionally, the proper level of assessment is that of individuals instead of "average programmer at the company."

The evaluation of cooperation related factors needs adaptation, as well (multisite coordination, stakeholder team cohesion). These aspects cover namely two kinds of teams: Intra-team communication at the unit performing ISVV where the interpretation is identical with that in the design organization; inter-team communication between the design entity and the ISVV site. A good approximate solution is a simple logic taking a pessimistic approach by assigning the minimum score of the two particular scores to these two communication-related aspects.

6.2.6 Motivating Example: Testing

One of the core issues is the dependency of ISVV related costs on the quality of the input artifacts (like software source code), while the output of the process has to comply with the strict quality requirements formulated in the standards.

This dichotomy in the requirements has non-trivial consequences on cost estimation.

- Testing of non-critical applications has as input typically a moderate quality code. It has to reach an acceptable, but not an extremely high fault coverage.
- Testing of safety-critical applications however, always targets high-fault coverage.

According to numerous observations, the detection rate drastically drops along the progress of the testing process after an initial peak (the testing time to detection rate distribution usually corresponds to a long-tailed Rayleigh distribution as shown in Figure 6.3, see [14]).

- This way effort estimation (the time to finish testing by reaching the required coverage) has to cope in case non-critical applications dominantly with the initial bulk of faults (which is relatively easy to estimate).
- Regarding safety-critical objects under test this termination time instance lies on the long flat tail resulting in a potentially large estimation error.

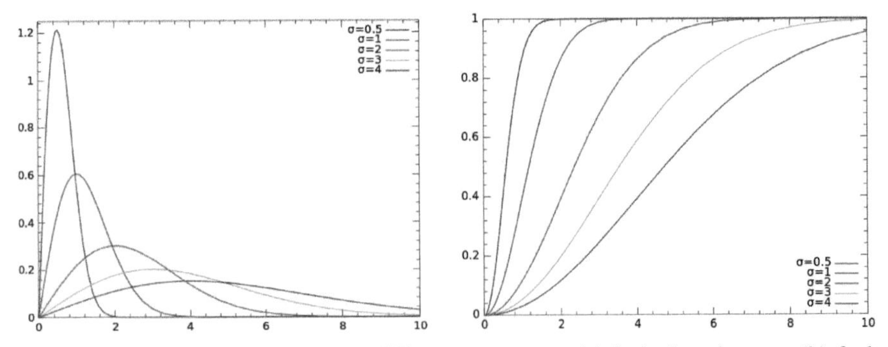

Figure 6.3 Rayleigh distribution by different parameters (a) fault detection rate (b) fault coverage (source: wikipedia.org).

This way the scoring scheme of traditional CEs is inappropriate, as a score corresponding to a "High input quality" would mean, that fault coverage is above a threshold. However, this corresponds in testing time to an interval between the threshold and infinity.

However, input characterization may improve estimation accuracy by substituting the original grade-like scores with continuous metrics. We used mainly visual Exploratory Data Analysis methods in order to find and validate main correlations and interdependencies on project data. During CECRIS, some typical and "atypical' projects were examined in order to see what connections and measureable characteristics can be set up between outputs and inputs of ISVV activities.

6.3 Experimental Results

Even the initial experiments delivered some highly valuable conclusions for prioritizing the individual goals in CECRIS.

Note that the low number of cases presented here is apparently insufficient to draw any conclusions with a sound mathematical foundation but they serve as an orientation for the further experiments.

6.3.1 Faithfulness of the Results

Figure 6.4 shows the effort estimated by COSYSMO compared to the real effort spent by the V&V activities for each pilot project.

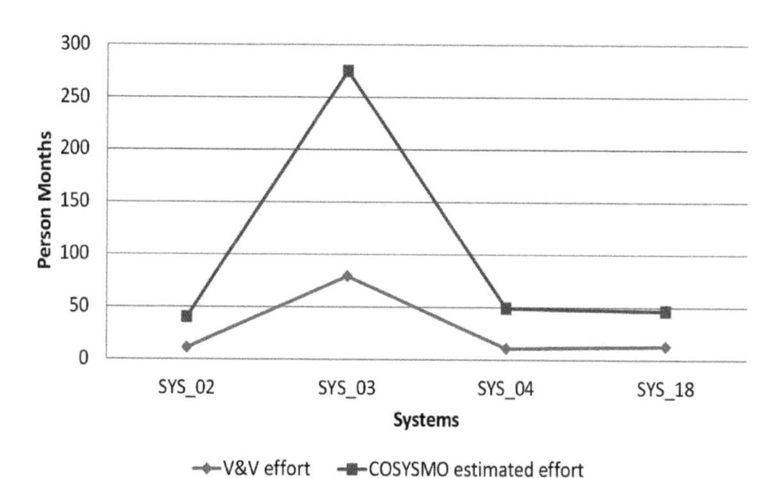

Figure 6.4 COSYSMO estimation compared to real V&V effort.

The ratio of the logged and predicted efforts is for the entire four projects $4 \pm 10\%$. The correlation coefficient is 0.99, which means a strict proportionality. (It is well visible in the figure that the two curves have a similar shape.)

This way, the first impression is that the direct application of COSYSMO to pure V&V phases delivers a qualitatively correct, but in absolute numbers wrong estimation. The main reason is that COSYSMO accounts for the efforts of the entire development process (starting with requirement analysis and including the entire code development and in-production testing phases, as well). The reference values from project logs cover only a smaller set of sub-activities of this process, V&V, accounting for 22–28% of the global project efforts.

The ratio of V&V efforts w.r.t. to global efforts is in a narrow interval over all the pilot projects.

Expert's estimations predict in general 30–35% share for V&V in the industrial practice depending on the project and product in contrary to the observed ~25% in the pilot calibration set. This better than average productivity originates in the very highly skilled and experienced personal typical in university close spin-offs.

Note that while the observed ~25% holds in the pilot calibration set, but this value should be further validated on a larger data set. It presents a further problem that the independent V&V and certification organizations have no exact data on the ratio to the total effort, while CECRISMO explicitly estimates effort for these phases.

COSYSMO can serve this way as a comparison base if the ratio of a particular V&V and certification activity can be properly approximated.

V&V efforts can be taken as a constant fraction of the predicted total development costs if only a rough estimate is targeted. Some COSYSMO implementations rely on the assumption on a constant ratio, which is valid if the entire project is carried out completely in-house.

However; the assumption has globally no proper justification in the case of an external accessor. The total effort estimation mixes activities carried out at the product manufacturer (3/4 share) and V&V carried out at the independent accessor. A rigid subdivision cannot properly predict V&V related efforts in general, as all factors may differ in the two actor enterprises.

However; such a first estimate is a candidate for a rough relative comparison of two solutions of V&V tasks carried out by the same company.

6.3.2 Sensitivity Analysis

Sensitivity analysis is widely used in order to estimate the impact of changing some input value or values onto the output of a function or a system.

The corresponding inputs are the size drivers and the effort multipliers in the case of cost estimation.

Sensitivity analysis forms the basis of answering numerous critical questions:

- **Impact prediction for process changes**: Process improvement necessitates always investments like the introduction of new tools into **technology**, improvement of team cohesion by teamwork support or investment into the human capital by training. While it is generally true that such investments has a beneficial impact onto the productivity and cost, *sensitivity calculation is able to predict their impact in quantitative terms*, as well.
- **Adaptive project management**: Sensitivity to the size drivers is a main characteristic to describe scalability of a project. Here the breakdown of the different size drivers into qualitative categories and taking *the individual sensitivity values for instance delivers a good indicator of system specification modifications during the process*.
- **Estimation of the impacts of misscoring**: The different effort multipliers need a special care from the point of view of sensitivity analysis. These are all categorical variables with assigned ordinals as domain (the effort multipliers take their score values from an ordered enumerated list).

By the very nature of the scoring by an expert judgment guided by an informal description is somewhat subjective.

The evaluator can score a particular factor by one score up or down. *Sensitivity analysis here answers the question what are the impacts if the evaluation expert puts a score to a wrong value* (typically some of the neighboring values of the proper one).

The estimation of the impacts of such scoring errors assures that a range of uncertainty can be provided in addition to the single effort estimate as the final result. (Some cost estimation methods apply a non-deterministic simulation around the expected scoring to deliver such an uncertainty estimate.)

The V&V focus of CECRIS necessitated this analysis, as the focus of the original COSYSMO description was not completely identical with the CECRIS objectives; this way, miscategorizations may occur. Another problem was that the scope of COSYSMO is essentially wider than that of CECRIS. This way, a significant part of the large domain of effort multipliers is irrelevant in the cases of CECRIS (for instance, no staff of very low skills and novice to certification will be involved into V&V activities despite the fact that a this is a valid score allocation in COSYSMO).

Accordingly; only a subdomain of the COSYSMO's score space is relevant for CECRIS and this raises the question of applying a narrower but finer granular set of candidate scores.

A sensitivity analysis of all the individual cost drivers was performed in order to estimate the potential impact of miscategorization and a refined scoring system.

This is basically performed by taking the nearest lower and higher score of an input factor. If the mismatch between the observed value and the predicted one drastically changes, thus the cost is highly sensitive to the particular factor than the scoring rules have to be revised.

If the observed values typically lie between the two predictor values corresponding to neighboring score settings, than a refinement of the input scoring (introduction of an intermediate value) may increase the resolution and accuracy.

The Figure 6.5 shows the relative impact of three cost drivers, **Requirement Understanding**, **Architecture Understanding,** and **Personnel Experience/Continuity**, on the cost estimator. Only these three cost drivers are analyzed for the sake of simplicity and without lose generality. As shown in Figure 6.5, increasing cost driver scores (x-axis), decreases the cost estimate (y-axis) nearly in a linear way. **Requirements Understanding** has a higher impact on the cost like **Architecture Understanding** and **Personnel Experience/Continuity**.

Even this small example indicates the importance of sensitivity analysis from the point of view of project- and process management.

If the quality of the requirement set specification is improved by a score than $\sim$20% can be reached in the terms of cost saving. If for instance, the traditional textual specification is substituted with an unambiguous formal model in an easy to understand form like the mathematically precise but well readable controlled natural language, then a significant cost saving can be reached with relatively low effort in training.

Naturally, multiple cost factors change in the case of the introduction of a new technology.

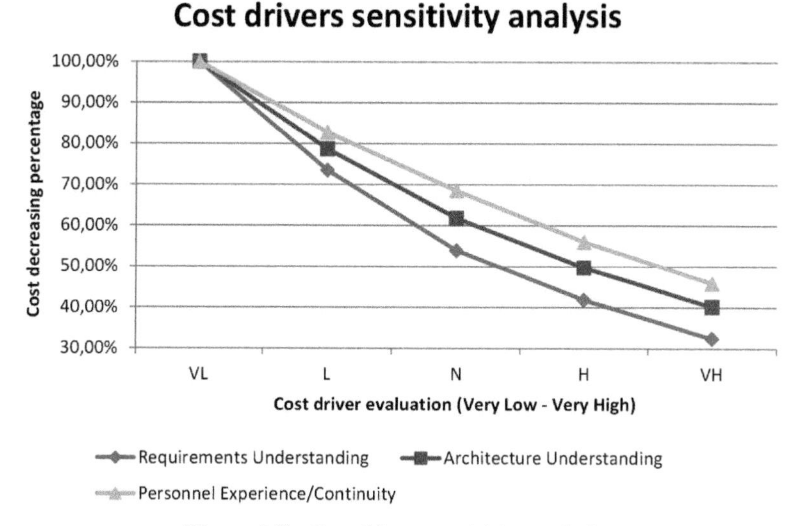

Figure 6.5 Cost drivers sensitivity analysis.

6.3.3 Pilot Use Case for Project Management

As pilot case for what-if analysis the core CECRIS action line was selected: In a company dealing with V&V/certification of critical ES advanced academic approaches are introduced. The skills of the local personnel would be at the beginning moderate and may only gradually reach the level of professional expertise. As a special case we investigated what happens if intensive coaching was provided by senior experts to help the transition. One of the previously analyzed projects was selected to carry out the what-if analysis.

COSYSMO calculator was taken and factors were determined as summarized by Table 6.1. Concentrating only on factors (taking values from Very Low through Nominal to Very High) where significant impact is expected, leaving other multipliers unchanged.

Our pilot calculation showed that a project introducing formal methods without experience ("Introductory") has approximately the same cost as a

Table 6.1 Pilot use case for introducing formal methods in verification

Factors	Orig.	Introductory	Final	Intermediate
Requirements understanding	H	N	VH	VH
Personal experience	N	L	N	N
Tool support	VL	H	VH	H

H, high; L, low; N, nominal; VH, very high.

project working with traditional, mostly human checks. Due to increased uncertainty, real costs (and technology related risk) may even be higher in this case. On the other hand, a guided "coaching" may result almost the same cost/effort saving as the ideal "final" stage, due to the reduced risk of novel technology and lack of experience (50 vs. 60%). Although these factors were not primarily calibrated for V&V (nor for embedded systems), such overall project management considerations also hold in this area.

6.4 Case Studies

In this chapter we present case studies to illustrate how the estimation method can be used on real-life data. During this analysis, we were also validating our assumptions about the process model.

6.4.1 Complexity Factors

Complexity is an essential factor, used both in COSYSMO and in *ad-hoc* effort estimations for ISVV. As we described in Section 6.1, one overall scoring for projects cannot describe the overall difficulty of the problem, therefore a more fine-grained estimation is needed. Scaling/rating of the input may also depend on the problem or domain, so this estimation should be calibrated to the domain and the nature of the problem as well (e.g., boot software, a communication submodule or a computationally intensive software component may be different). Note that this complexity is expressed also in the COCOMO family as "algorithmic complexity". In case of ISVV, we can rely on the advantage that the software to be checked is already available, so metrics which are typically measurable on the outcome of a traditional development project can be used for input characterization.

Although the interpretation and calculation of complexity differs across domains and development steps, we took some examples which are easy to calculate. Here we introduce complexity in the sense of code complexity.

Code complexity is measured usually on the source code of components. It concentrates on measuring the structure of the code (which do not necessarily correlates with the actual complexity of the problem solved by the particular software component, which is the main criticism wrt. complexity measurements).

One of the most widely used metrics is McCabe complexity:

$$\#of\ linearly\ independent\ paths\ in\ CFG =$$
$$\#Edges - \#Nodes + 2 \times *\#ConnectedComponents$$

This metric fits well with problems which are control dominated, captured by the structure of their Control Flow Graph (CFG).

The above definition is independent from the programming language but is influenced by the coding convention, therefore its application may be limited for domains/teams. It should be quantized (e.g. low-medium-big). A typical upper threshold for "low" complexity may be five for instance in embedded software.

Another important metric is the *Halstead* metric concentrating on the "language" of the software, measured in the number of terms used.

Halstead metrics are concentrating on the "size" of the software measured in the number of "operators" and "operands" in software.

Let n1 be the number of *unique operators* and n2 the number of *unique operands*, while N1 denotes the *total* number of *operators* and N2 stands for *total* number of *operands*.

$$\textit{Vocabulary}(n) = n1 + n2$$
$$\textit{Length}(N) = N1 + N2$$
$$\textit{Volume}(V) = N * \log_2 n$$
$$\textit{Difficulty}(D) = (n1/2)*(N2/n2)$$
$$\textit{Effort}(E) = D*V$$
$$\textit{Delivered Bugs}(B) = V/3000$$

It is important to note that the latter two derived metrics are highly domain-dependent. Effort refers to actual effort creating the software while "Delivered Bugs" is a rough estimation of "faults inserted". E.g., the number of Delivered bugs is known to be an underestimate in general-purpose C/C++ programs but is in the range of ISVV found bugs, since ISVV deals with "residual" bugs not detected by in-house testing and other means of fault detection.

The programming language itself has a trivial impact on these metrics, especially the Halstead length of a program.

These metrics can also be applied at the model level since most of their concepts are already captured by (detailed) software models. Similarly to McCabe metrics, Halstead (derived) metrics can be used to calculate size drivers for COCOMO-like estimations.

We have recently experimented (among others) with C, Java, Python, JavaScript and found that these metrics may be used on only for input characterization, but also for finding outliers. For instance, in the examined subset, there were different solutions available for a computationally intensive problem written in the same programming language with drastically different complexity metrics. It turned out that one naïve solution is inefficient, but simple iteration logic while the other solution relied on specialties of the problem. This second component would be trivially harder to maintain but runs with less execution time for the majority of the inputs. However, this is such a difference which could not have been found only by looking at the requirements or test cases of the given software component.

These examples also illustrate that the COCOMO family is too high granular when talking about factors, basic estimations should be supported by more detailed metrics. Most of the effort multipliers, however, will remain similar. These methods can also be used to select the appropriate level and target of the application of (semi) automated formal methods.

6.4.2 Cost Impact of Requirement Management

Taking the example of a breakdown of a difficult requirement to five simple ones, the difference is 50% (2.5 vs. 5) in nominal difficulty. Of course, these numbers represent more rules of thumb than exact calculation, but still they express industrial best practice in project planning. Structuring of requirements is therefore a crucial part of project design. Although this can be done in any textual tool or in Excel (the most widespread requirement definition tool), a structured, object-oriented approach can help requirement refactoring and reuse by introducing typed interdependencies and domain specific notation in requirement specification.

As requirements have a specific "lifecycle", changes and modifications have significant impact on overall project difficulty, and therefore affect related ISVV activities.

Table 6.2 shows, for instance, that a requirement of "Nominal" difficulty has a 30% less effort implications if it is reused from an existing requirement set.

Table 6.2 Effect of requirement lifecycle

No. of System Requirements	Easy	Nominal	Diff.
New	0.5	1.0	5.0
Design for Reuse	0.7	1.4	6.9
Modified	0.3	0.7	3.3
Deleted	0.3	0.5	2.6
Adopted	0.2	0.4	2.2
Managed	0.1	0.2	0.8

Requirements here are measured in a uniform way, however, from safety critical point of view, there can be huge differences even between requirements of same category (e.g., change in a "nominal" requirement related to emergency shutdown may have larger impact on software testing).

Also there are interdependencies among the requirement set, which may override the above numbers.

6.4.3 Automated Analysis for Factor Selection

Besides exploratory analysis methods, a wide selection of automated analysis technique is available in "algorithm as a service tools" like IBM Watson Analytics, Microsoft Cortana or other evolving services. These tools typically support data cleansing and evaluation by combining a selection of well-known statistical algorithms with heuristics and other (e.g., text mining based) methods in order to find correspondences between input variables of datasets.

These tools return with a number of suggestions which in turn can be justified, refined (or even invalidated) by a profound analysis and check of human experts. These automated methods can not only speed up the initial steps of data analysis, but also systematically reduce the chance of overlooking factors or biasing analysis by false assumptions.

We analyzed 7 different ISVV projects with source code files in the range of 500 from the aerospace domain and submitted input information (complexity measurements, number of requirements/files to check, etc.) and output evaluation (number of RIDs found, information about these RIDs w.r.t. input artifacts, overall effort estimation) and performed automated analysis in order to see what correspondences and implications are derived. We were using IBM Watson Analytics. Our findings were the following:

Input quality. As expected, data quality was very heterogeneous across projects. The tool pointed out factors where the number of missing entries (NAs) or constant data may distort the analysis. Input quality information is important as filtering out irrelevant or partially missing factors can prevent later analysis methods from generating false/not interpretable results.

Categorization. Without any previous domain knowledge, some interesting categorization suggestions (expressed in the form of automatically derived decision tree) were found. The categorization returned by the tool was approximately the same as returned by experts. Such information might be used in qualitative "labelling" of projects.

Heat map on frequent combination of factors. When selecting input factors, it is important to know which combination of values (intervals) appear in data to see how realistic/relevant the scaling of factors is. Analysis methods returned some important suggestions, for instance on how to combine readability metrics and file size to find a rough estimator for input quality (fault density measured in the “frequency” of RIDs).

Derived factors. The analysis tool was also able to derive estimators (based on partial linear regression) which can take the project characteristics into account and return function-like closed formula predictors where parametrization may depend on project nature (which may be derived from the same metrics as used for project categorization).

Of course, the above estimations/predictions do not hold for all V&V projects, but such automated analysis results may speed up the domain specific estimation and quality improvement process by inserting systematic analysis into today’s mostly ad-hoc method.

6.4.4 Quality Maintenance Across Project Phases

In order to see the effects of multiple-phase development (and, therefore, multiple-phase ISVV activities), we took two examples from the aerospace domain [9].

Our research questions were the following:

- How does the “coding style” (i.e., structuring the code in smaller/bigger files with corresponding “header/specification” information) affect the quality of the code?
- How do the number if iteration and the timespan of the project (and, therefore, the ISVV activities) influence the quality?

As Figure 6.1 shows, quality of artefacts may affect subsequent project phases, and thus, the input of the corresponding V&V activity. In critical projects, traces among phases are typically available and support the identification of engineering decisions (e.g., how many classes will implement a certain requirement).

We also tried to introduce metrics to support the evaluation of dependency between phases in a quantitative way; the number of requirements/KLOC is such a candidate.

Main findings of this experiment were the following:

- Although traceability is assured by the development process and tools, the effect of a fault inserted in an early phase (e.g., in a requirement) is not always corrected, especially during re-iteration.

- Mainly due to the mostly human work during verification phases, RIDs are often recorded in a way which does not help later analysis: typical faults (e.g., in coding style, comments, or even implementation not fully consistent with design documents) may be reported once but may refer to multiple artefacts. Moreover, this also underlines that the number of RIDs may not be a precise characterization of ISVV project output, since the granularity of RIDs might move on a wide scale.
- The timespan of the original project heavily influences the "lifecycle" of artifacts and also the quality assurance. We concentrated mostly on completeness and correctness faults, as during the re-iteration phase of ISVV, these get a special emphasis. We measured the fault ratio per KLOC in the source code. In the case of an "incremental" project with shorter iteration times (see the left part of Figure 6.6), the project converges and the ratio of faulty input items drastically decreases. In the case of long iterations and multiple changes in requirements, both the completeness and correctness of the software code got significantly worse. Note that these relatively small fault rates (when compared to a rule of thumb of 5 faults/KLOC) are detected after the code has been approved by in-house checks.

This experiment also underlines that measuring only the characteristics of input artefacts is not enough without the organizational and human factors,

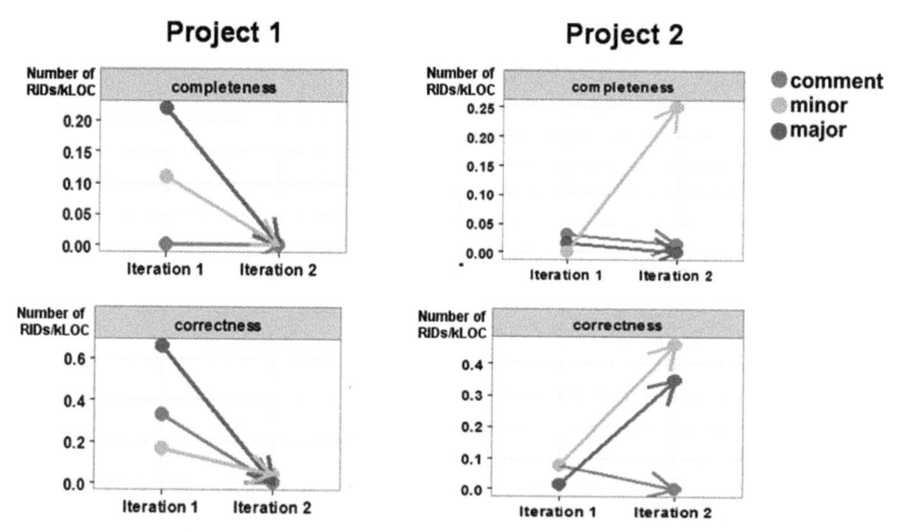

Figure 6.6 Trends of fault in multi-phased ISVV projects.

which obviously have a huge impact on project effort both at the customer side and at the ISVV company.

6.4.5 Fault Density and Input Complexity

In the following section, we present an analysis example which tries to set up high level correspondence between fault density (measured *post-mortem*, at the end of closed ISVV project phases) and complexity metrics (measurable at project planning stage).

A simplified and cleaned set of data has been loaded to Microsoft Power BI tool used for data analysis and visualization. Results presented here do not depend on the particular tool of choice. Besides basic “BI” visualizations, some R code was used to generate the plots. A part of the dashboard is shown on Figure 6.7.

The left figure shows the number of RIDs (indicated by the size of circles) and their correspondence with the complexity metrics (McCabe and maximum nesting). The figure suggests that there is a more direct influence of complexity on fault number than the maximum nesting value. The figure on the right shows the distribution of RIDs according to their severity. Besides other findings, our experiment also showed that there is a clear negative correlation between comment to code ratio and code complexity.

Some high level conclusions on RIDs found in this project:

1. **Comment** RIDS (which are often neglected by the customer, especially if a project is close to mission start) may have been found by ISVV or prevented during development if basic formal methods and automated, “template-like” check were used.
2. In the case of **Minor** and **Major** RIDs, requirement traceability is a key factor in efficient V&V. These result in more objective RIDs which are

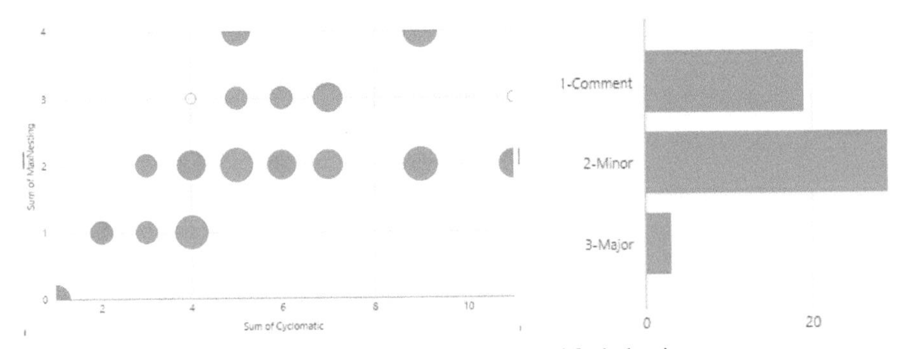

Figure 6.7 Complexity metrics and fault density.

also easier to validate and accept by the customer, while their correction can also be better traced.

Major RIDs may include a wide variety of discrepancies, from potential deadlocks caused by obsolete code to using wrong data types which may result in buffer overflow. Some of these faults are hard to find using manual methods, but are supported by formal tools which also return counter examples representing (potentially) wrong behavior. Major RIDs also require peer review, and are mostly accepted by the customers. Introducing Behavior-Driven Design technologies in the development may also help focusing testing and V&V effort and eliminating faults caused by inconsistent development phases.

6.5 Conclusions

Cost estimation is a fundamental pillar stone of all project management activities. Traditional systems and software engineering can rely on a variety of CEs providing sufficiently accurate predictors on the efforts needed to a particular application development. Independent systems verification and validation of safety critical applications is a crucial activity in assuring the compliance with the standards.

The current chapter evaluated the possibilities of creating a cost estimator dedicated to the V&V phase of system design. The creation of such an estimator is feasible currently primarily due to the unavailability of a sufficiently large calibration dataset. However, a proper adaptation of traditional software CEs has proven its usefulness in process improvements and what-if style evaluation on changes in the workflow.

The adaptation of software cost predictors is not a mechanical process. Differences in assigning a metrics to the complexity of the target project, the scoring of the highly skilled personnel, the impact analysis of the introduction of sophisticated tools, etc. all need a domain expert.

As ISVV is a follow-up activity of a design phase, the design quality heavily influences the amount of work to its execution. The quality of outputs of the previous activities (e.g., design in the V&V Plan, Requirement Verification, Hazard Analysis) are major Effort drivers for rest of the V&V cycle. Modeling this behavior will also allow to perform ROI analyses at early stage of the V&V process and/or continuously monitoring the cost-quality tradeoff of the overall V&V cycle.

A major innovation of the approach is the use of advanced exploratory data analysis techniques [15] to get deep insights into the ISVV process.

Finally, the chapter pinpoints that minimal dataset which is a recommended target of project logging to support future process improvement.

References

[1] CECRIS. *FP7-PEOPLE-IAPP-CECRIS-324334 D2.1 Assessment methodology for analysis of companies V&V processes.*

[2] CECRIS. *FP7-PEOPLE-IAPP-CECRIS-324334 D2.5 Definition of Certification Oriented V&V Processes.*

[3] CENELEC. (1999). *EN 50126: Railway Applications – The Specification and Demonstration of Reliability, Availability, Maintainability and Safety (RAMS).* Brussels: CENELEC.

[4] CENELEC. (2011). *EN 50128: Railway applications – Communication, signalling and processing systems – Software for railway control and protection systems.* Brussels: CENELEC.

[5] CENELEC. (2003). *EN 50129: Railway applications – Communication, signalling and processing systems – Safety related electronic systems for signaling.* Brussels: CENELEC.

[6] ISO. (2011). *ISO26262 – Road vehicles – functional safety, International Organization for Standardization.*

[7] ISO (International Organisation for Standardisation) and IEC (International Electrotechnical Commission). (2009). "Software Product Quality," in *ISO/IEC 9126*, November 2009.

[8] RTCA. (2012). *RTCA/DO-178C, Software Considerations in Airborne Systems and Equipment Certification.*

[9] Pataricza, A., Gönczy, L., Brancati, F., Moreira, F., Silva, N., Esposito, R., Salánki, Á., Bondavalli A. (2016). "Towards an analysis framework for cost & quality estimation of V&V project," in *DASIA 2016*, Tallin, Estonia.

[10] Brancati, F., Pataricza, A., Silva, N., Hegedüs, Á., Gönczy, L., Bondavalli, A., and Esposito, R. (2015). "Cost Prediction for V&V and Certification Processes," in *2015 IEEE International Conference on Dependable Systems and Networks Workshops (DSN-W)* (New York, NY: IEEE), 57–62.

[11] COCOMO. (2015). *COCOMO II – Constructive Cost Model.* University of Southern California.

[12] Constructive System Engineering Cost Model (COSYSMO). Available at: http://cosysmo.mit.edu/

[13] Fortune, J., Valerdi, R., Boehm, B. W., and Stan Settles, F. (2009). "*Estimating Systems Engineering Reuse*." MITLibraries. Available at: http://dspace.mit.edu/handle/1721.1/84088.

[14] Kan, S. H. (2002). *Metrics and Models in Software Quality Engineering*, 2nd ed. Boston, MA: Addison-Wesley Longman Publishing Co., Inc.

[15] Pataricza A., Kocsis I., Salánki Á., Gönczy L. (2013) "Empirical Assessment of Resilience," in *Software Engineering for Resilient Systems. SERENE 2013*, eds A. Gorbenko, A. Romanovsky, V. Kharchenko. Lecture Notes in Computer Science, vol. 8166. Springer, Berlin, Heidelberg.

7

Lightweight Formal Analysis of Requirements

András Pataricza[1], Imre Kocsis[1], Francesco Brancati[2], Lorenzo Vinerbi[2] and Andrea Bondavalli[3,4]

[1]Dept. of Measurement and Information Systems, Budapest University of Technology and Economics, Budapest, Hungary
[2]Resiltech s.r.l., Pontedera (PI), Italy
[3]Department of Mathematics and Informatics, University of Florence, Florence, Italy
[4]CINI-Consorzio Interuniversitario Nazionale per l'Informatica-University of Florence, Florence, Italy

Requirements are the core work items of the design and checking workflow target safety critical systems. Accordingly, their completeness, compliance with the standards and understandability is a dominant factor in the subsequent steps. Requirements review is a special kind of Independent Software/Systems Verification and Validation (ISVV). The current chapter presents methodologies to use lightweight formal methods supporting experts in a peer review based ISVV.

7.1 Introduction

The quality of requirements dominates the efforts of a design process especially in the case of safety-critical applications (see Chapter 6). As described in the previous chapter in details, the effort and quality of the ISVV heavily depend on the input quality of the work items submitted for review by the customer. Frequently a significant part of the efforts is wasted to basic activities similar to the data cleansing phase in the field of data analysis regarding their level and ratio (which can reach a few tens of percents). For instance, ill-structured documents, inconsequent and non-conformant with the standards use of terminology all require expert effort to be checked although they do not form the essence of the assessment.

Correspondingly, the exhaustiveness of the checks performed has a major impact on all the activities relying on the completeness, standards compliance and integrity of them. This way, requirement review has to be as thoroughgoing as possible. However, this part of the workflow benefits only to a moderate extent of the advantages of Model-Based System Engineering (MBSE) and formal methods due to the typically conservative (informal) or at most semi-structured text based formulation used mostly in the industry.

The objective of the current chapter is the presentation of an approach targeting a gradual introduction of MBSE and formal methods to requirement checking. The introduction of easy-to-use methods simultaneously assures an increased productivity and quality without the need of a single step introduction of a complete framework or specialized skills in formal methods.

The chapter introduces the basic modeling concepts as defined by standards. The subsequent section presents techniques carrying out extended syntactic analysis over the requirement documents. After addressing change management in iterative requirement design/modification-checking workflows the closing section deals with the integration of the measures described into the ISVV.

7.2 Objective

Our objective is supporting dominantly peer review based ISVV executed by SMEs. Typically, highly-qualified experts constitute the personal of such companies. Reviewers usually are very familiar and knowledgeable of the application domain without deep skills in advanced formal methods.

Accordingly, our evolutionary approach is less ambitious, than a revolutionary one exploiting the full potential of mathematical proof of correctness methods. It follows the paradigm of hidden formal methods in which the user of the tool gets the support from a built-in intelligence, but the working environment has no or very moderate changes compared to the traditional one.

Our approach does assume either an end-to-end MBSE or a complete automation of the workflow; however, it can contribute to a significant effort saving by reducing the overhead originating in document cleaning, managing the progress of the assessment including checking its completeness.

This way, the efforts of the experts can be focused on the hardcore problems related to technical evaluation deliberated of the majority of the pure mechanical tasks not requiring their expertise.

The subsequent sections address the three major questions in improving ISVV:

- How to create interchangeable and well-structured documents out of traditional unstructured ones?
- How to create a domain-specific working environment out of a traditional one by adding quality improvement and checking measures based on hidden formal methods?
- How to improve the convergence of requirements by change management in iterative design-ISVV workflows?

7.3 ReqIF and Modeling

The requirements play an important role in cooperating between final product manufacturers and part suppliers, as these provide the basis of outsourcing and acceptance tests between them. This way, the exchange of requirement between cooperating partners is a crucial part for instance in the automotive industry demanding relatively low interaction times.

The Object Management Group (OMG) developed an open standard [1] called Requirements Interchange Format (ReqIF) to assure interoperability between cooperating partners (see Figure 7.1). This standard is open toward different design and checking technologies thus it is a natural candidate to information exchange between designer and independent software/system

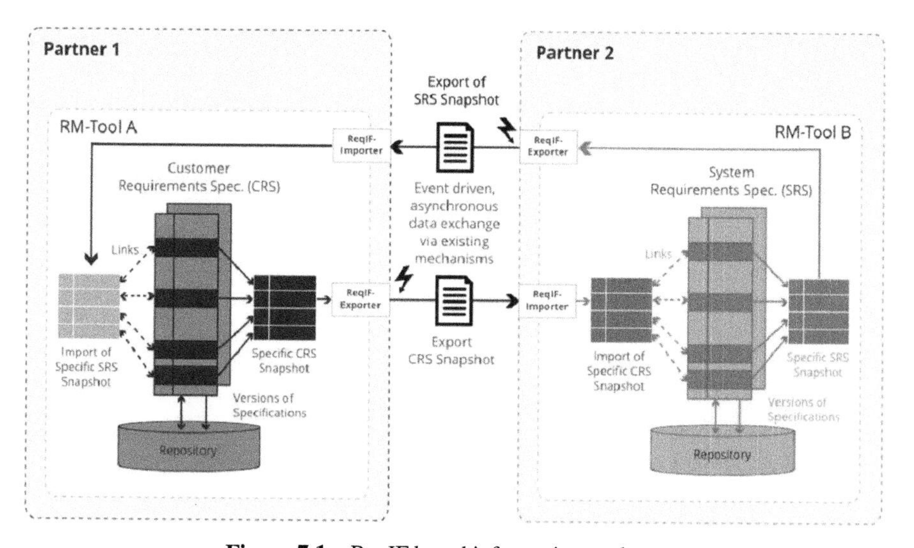

Figure 7.1 ReqIF based information exchange.

verification and validation (ISVV). OMG ReqIF provides a well-regulated set of rules and protocols for cooperation.

Requirements Interchange Format has wide support regarding open and commercial requirement design and management tools. Also, leading vendors put requirements as the core entity of the entire design and checking work-flow. Advanced MBSE based frameworks integrate requirement management with design and test. Traceability is a priority concept in ReqIF.

The following summary presents the main benefits of ReqIF as an exchange model language for ISVV based on the top-level constructs in its meta-model.

Requirements Interchange Format supports the exchange of core content labeled by a header sufficiently detailed to identify the document itself and optionally tool specific extensions from other information sources, like results of evaluation (Figure 7.2).

The specification is the core content of a ReqIF instance associated with specification types, objects and the relations between them (Figure 7.3). The notion of links (called here "SpecRelations") assures the traceability of the requirement model to other artifacts. The metamodel supports hierarchical composition of requirement sets, like a well-structured description of safety cases derived by specialization from a standard, e.g., as an evidence list and interlinked supportive arguments.

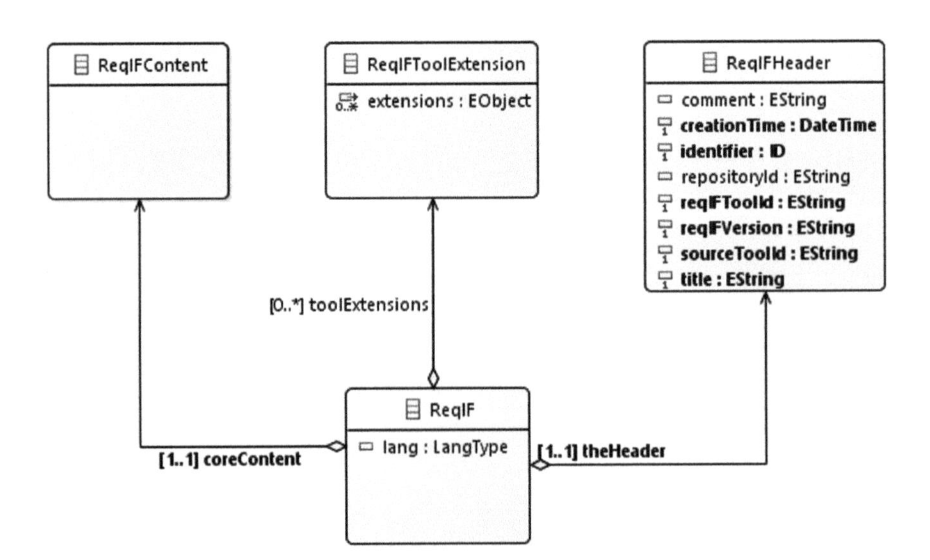

Figure 7.2 Exchange document structure.

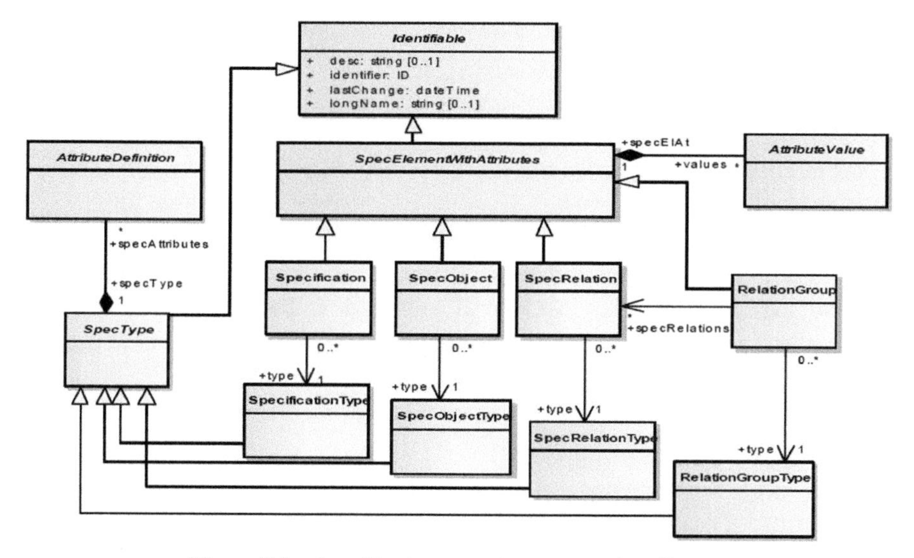

Figure 7.3 Specifications, requirements, and attributes.

The most popular infrastructure for ReqIF is Eclipse RMF (Requirement Management Framework) [2]. ProR [3] an open-source editor to edit structured requirement documents.

Definition of data types, including enumeration types, supports the creation of a domain-specific MBSE-styled model representation.

For instance, the different Safety Integrity Levels form a core enumeration datatype of the form of $\{SIL0 \ldots SIL4\}$. Traditional type checking assures the avoidance of omissions or ill-specified values in the corresponding field.

Syntax-driven editors constrain the designer to use only such values in the field, that comply with the datatype definition.

However, turning a column in an Excel worksheet from the general string type to one out of the predefined values forming the enumerated data type implements simply the same principle. Such constraints simply prohibit entering a wrong value into the instance model corresponding to the application under development.

Moreover, this tiny example indicates a further opportunity in separating the duties: If a domain and modeling specialist designs the Excel template, he could embed the terminology, structure, etc., from the standards, as well. The application designer filling out the template with the content corresponding to the particular application under development will face his traditional working environment with the hidden type model and check already embedded by the expert.

At the same time, the example pinpoints the limitations of a pure ReqIF-based working environment design approach, as well. Implementation of complex relations necessitates low-level (e.g., VisualBasic) programming and it benefits of modeling only by starting from a proper blueprint, which implies all the drawbacks of traditional programming.

Bidirectional communication between cooperating partners was a primary design objective of the OMG ReqIF standard. In the context of ISVV, this offers the opportunity of using it in the ISVV-to-developer communication for feeding back the review results in an entirely standards compliant way. This way, the iterative process can benefit from the rich navigation and traceability supporting features of ReqIf.

7.3.1 Domain Conceptualization

The industrial success of ReqIF in the inter-party communication in product design makes it a natural candidate in developer-to-assessor cooperation and model-based ISVV, as well. Moreover, as ReqIF documents carry both the instance model and its respective metamodel, they can harmonize of requirement design and ISVV.

At the top end, ReqIF-based requirement modeling serves as the starting point of sophisticated methodologies aiming at correctness by design (like RODIN – Rigorous Open Development Environment for Complex Systems [12] transforming specifications into formal Event-B models). However, the introduction of heavyweight formal methods into ISVV, a single phase of the product development process faces serious obstacles regarding skills, and in the overwhelming majority of ISVV tasks, it has an improper modeling effort/benefit ratio.

Our approach uses ReqIF similarly for information exchange, as this assures a well-structured requirement set. Lightweight modeling should complement the methodology of customization of the ReqIF metamodel and work environment of the requirement composer to a particular product or product family using MBSE. Finally, the customized work environment accommodates traditional, manual, design, and V&V methods, as well.

Ontologies serve as primary candidates for semantics based unification and conceptually clean metamodel design [4]. Ontologies are formalized vocabularies of terms covering a specific domain. They define the meaning of terms by describing their relationships with other terms in the ontology. They classify the terms that can be used in a particular application, characterize possible relationships, and define possible constraints on their use by providing formal naming and definition of the types, properties, and interrelationships [5].

Knowledge organization, complexity reduction, and problem solution all use ontologies for a variety of fields ranging from the Semantic Web, through systems and software engineering to such non-technical fields, as library science. The main use case of ontologies is conceptual data integration.

The driving force behind their standardization of formats (RDF and RDF Schemas, OWL) is the World Wide Web Consortium (W3C). The formats support interoperability, information fusion, and interchange.

MBSE largely depends on metamodeling (UML and derivatives). Metamodeling and ontologies are two different, but mutually transformable approaches[1] to modeling language and model construction. Both paradigms focus on the description of the relations between concepts, checking of the compliance of instances (individual models) with their respective parent metamodel or upper ontology.

In contrary of usual metamodels, ontologies have a precise semantics regarding mathematical logic, for instance in ISO/IEC Common Logic [6]. Ontology tools have built-in functions checking the completeness and consistency of the models, and the correspondence of subontologies (specializations) and instances to their upper ontology (subsumption check).

The gradual introduction of hierarchical and relational elements into the model following a vocabulary–taxonomy–ontology process results in an ontology corresponding to a particular standard. Such an ISVV ontology consolidates notions and their mutual relations defined in standards as concepts.

Ontology processing has supportive mechanisms for information fusion by virtually merging multiple, physically separate ontologies. Starting from multiple ontologies representing different viewpoints facilitates aspect-oriented modeling.

[1]Theoretically, not all ontologies have an explicit metamodel counterpart, but the subclass of ontologies referred in the current chapter is subject of metamodeling based design. For instance, the Object Management Group (OMG) offers a bridge in the form of an "*Ontology Definition Metamodel*" [13].

The requirement set related to a particular application (legacy documentation, source code, and comments, etc.) are then instances of this ontology. This way the interdependence of entities and V&V steps managing them is explicit.

7.3.2 Integration with Existing Practice of ISVV

A (slightly simplified and obfuscated) real-life example taken from a railway hazard analysis project serves as motivating example.

The railway is a safety-critical domain; various safety measures designed into the system address the hazards that pose an unacceptable risk; these have to be proven to mitigate the various risks to an acceptable level.

The assignment of so-called Safety Integrity Levels ranging from 0 to 4 classify safety instrumented systems and functions. Each level has an associated interval of probability of failure on demand of the safety function, what translates to an overall risk reduction capability.

Designers of the original documentation used a plain, unstructured list of hazards as the input for risk analysis (Figure 7.4). However, structuring the potential causes indicates clearly its flaws. At first, the introduction of the abstract concepts "*Subject*" and "*Impact*" separates the different aspects related to a hazard event (Figure 7.4).

Aspect weaving in the form of interrelating them derives the individual categories, like "*Line Controller Death*." The inclusion of "*No Hazard Event*" and "*Fire*" do not fit into the scheme. The list of hazards is still incomplete w.r.t cardinality constraints. For instance, the standard may require the complete coverage of all potential hazard events by evaluating all "*Subject*" and "*Impact*" combinations. Automated reasoning reveals that the "*Staff OnBoard Death*" category lacks the considerations.

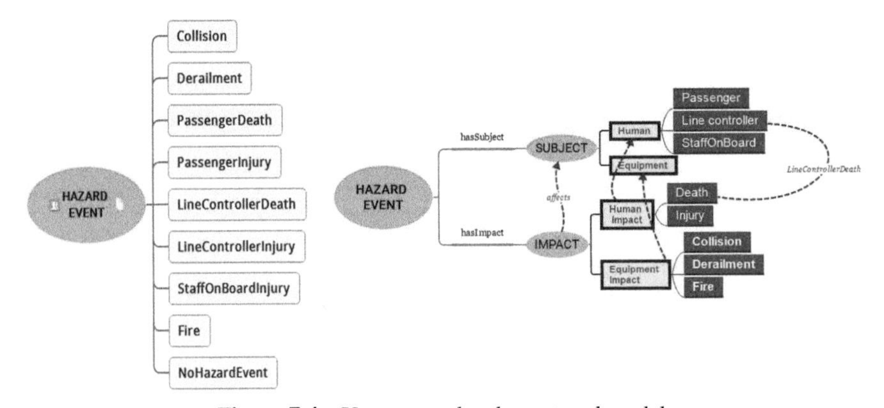

Figure 7.4 Unstructured and structured model.

At the same time, this model is easy to maintain. For instance, after the introduction of the notion of a "*Driver*" as a separate category, inherence mechanisms can derive the two subcases "*Driver Death*" and "*Driver Injury*" without touching other parts of the model.

Moreover, the design and ISVV workflows may rely on external information sources, as well. Information fusion necessitates the unification of the concepts of the different data sources by establishing the correspondence between their notions.

For instance, risk analysis should cover all the hazards above a given frequency of occurrence, which necessitates the inclusion of historical statistical data from external data sources (like [14] in Figure 7.5). Their integration into the ontology can follow the same unification approach, as in [7] based on mapping the notions in different models after some elementary operation (like calculating totals when aggregating overly fine granular statistical data).

Note, that the process of information fusion is an important engineering task and not a pure semantic matching of two models. Apparently, resolution of the two models differ merging different categories in the statistics into a single concept in the model of hazard events assumes a similarity in their occurrence and impacts. Aggregation of categories is at the same time an input specification for the underlying summation of frequencies of their occurrence.

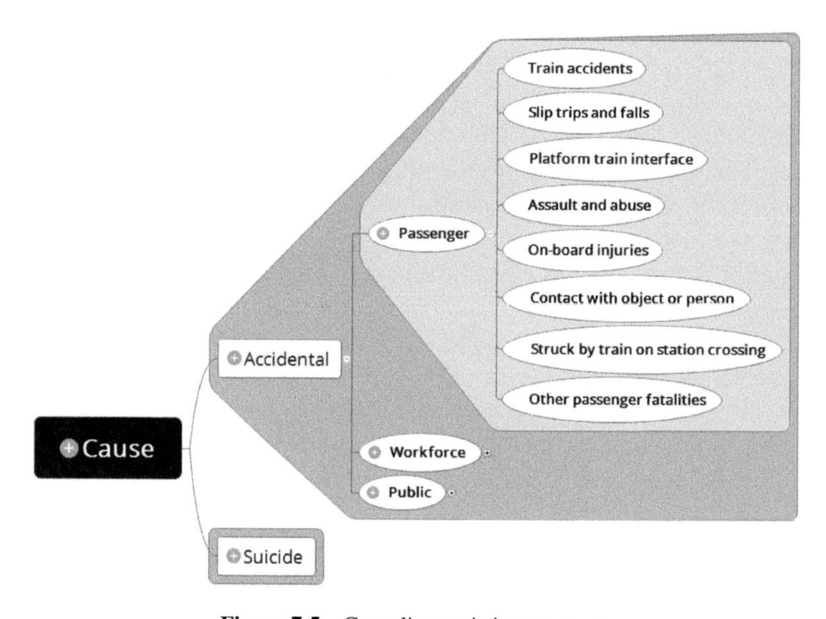

Figure 7.5 Causality statistics structure.

Such an interrelation of statistical data and the input model of hazard analysis support augmentative maintenance of the model. The appearance of a new category in the statistics (e.g., security) with no counterpart in the hazard event ontology pinpoints that the later one is not up-to-date.

7.4 Requirement Change Propagation

Our motivational example comes from the railway domain loosely based on an actual change scenario, similarly as the example above. It highlights the importance of lightweight formal methods from a further aspect, change management. For didactical as well as legal reasons, the case presented here is very heavily simplified and sanitized from multiple aspects.

The SIL of a function has a fundamental impact on its development cost and time, as higher levels require increasingly sophisticated V&V activities. Consequently, during requirement change impact analysis it is essential to correctly identify whether a requirement change indirectly causes SIL changes in a specification through change propagation.

7.4.1 Original Specification

Our example demonstrates how the changes in the requirement set of a Central Traffic Control system have a propagation effect in the whole specification.

Keeping station area traffic safe is a complex problem involving many tracks and switches in a complex manner, and the risks stemming from a significant number of hazardous situations have to be mitigated. Trackside signals regulate station area traffic allowing or denying entry to a track or (switching) point. Classically, the control of the traffic through the signals has been performed by local personnel and systems. The main means of risk mitigation is *signal interlocking*: a separate system overrides any traffic control command that would lead to a hazardous signal configuration. (For instance, giving a "clear" signal at the same time at two entry points of an interlocking.) Signal interlocking can be overridden in the local traffic control system under strict operational rules, e.g., a switch with broken switch state monitoring correctly halts traffic; however, to resume traffic, local personnel has the situational awareness and authority of a temporal override of the associated signal interlocking.

Traffic control has been and is being centralized worldwide To increase operational efficiency, a Central Traffic Control (CTC) system manages the traffic at multiple stations. CTC can be an overlay to the existing systems without substituting the local traffic control and the remote CTC "pushes its buttons" instead of local personnel. Local signal interlocking is left unchanged, too.

The "original" specification on Figure 7.6 represents a simplified excerpt from the specification of such a system. Importantly, the CTC does not have the full authority of local personnel; it is not allowed to issue signal

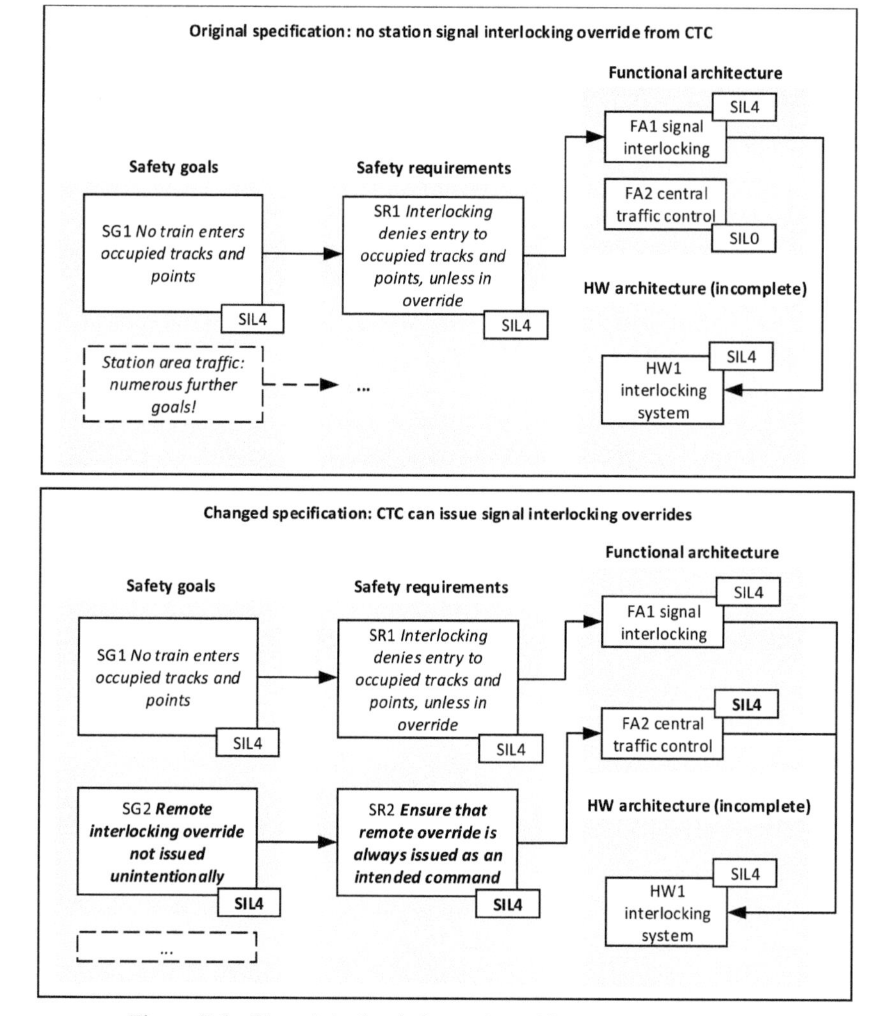

Figure 7.6 The original and changed specification in our example.

interlocking override (more generally, safety-critical) commands. As a result, the process of setting up safety goals, decomposing them into safety requirements and mapping those onto elements of the functional architecture identifies that it is SIL0 (not safety critical). On the other hand, notice how the high-risk mitigation capability requirement (SIL4) is carried over from the safety goal to the interlocking system.

7.4.2 Changed Specification

This specification has the chance to lead to a highly safe system that conforms to the legal requirements on safety. However, an operator is also concerned about operational efficiency. Let us assume that the operator finds the above specification too restrictive; in many circumstances, override situations can be managed acceptably safely, even if the override command is issued remotely. However, some characteristic hazards have to be avoided [8]; one of them is the CTC issuing override commands unintentionally. (The CTC is usually a complex, software-based system, where operators manage multiple stations of a geographical region with reduced situational awareness due to their remote location.) This leads to the specification excerpt depicted in Figure 7.6 as the changed specification.

The key difference between the two specifications from change impact analysis is that the central traffic control became a safety-critical component. Risk mitigation assumes the absence of override commands issued by the central traffic control unintentionally.[2] This change in SIL has further propagating effects; the V&V activities associated with the architecture, interfaces and implementing components of central traffic control have to be revisited.

7.4.3 The Change Impact Propagation Method

Requirement engineers have to evaluate the propagating effect of changes and rework the specification accordingly. This task involves two major phases.

- ***Suspicion marking through change impact propagation***. The directed dependency graph of the specification is traversed starting from the initial changes introduced into the specification. Dependencies and requirements that may have to be changed as an effect of the original change are marked as *SUSPICIOUS*. After that, specification objects

[2]There are many ways to ensure this, regarding the operators as well as the software/hardware system; discussing these is not in the scope of this chapter.

that are connected to *SUSPICIOUS* ones are evaluated and potentially marked, too in a transitive manner. Effectively, the *SUSPICIOUS* marking is "propagated" in the reachability subgraph of the originally changed elements. The resulting *change impact cover* – the subgraph defined by the vertices marked *SUSPICIOUS* – is passed on to marking processing.

- ***Processing marking***. One by one, the suspicion-marking of the marked dependencies and requirements has to be either accepted or refuted. If accepted, the appropriate specification change has to be designed and performed.

We are mainly concerned here with the first phase, although the value-based change impact propagation we introduce gives guidance to the second one, too.

- In practice, manually performing the first activity is a repetitive, time-consuming and error-prone task even for moderate size specifications.
- The best of breed modern requirement management tools support topology-based propagation: anything that is connected to a specification element marked *SUSPICOUS* is *SUSPICIOUS*, too.
- Some modern tools begin to support type-based propagation. In this case, marking is propagated only along the configured types of dependencies and only upon the configured types of requirement attributes becoming *SUSPICIOUS*.

Type-based propagation is a powerful tool to reduce the extent of the change impact cover in the specification. Observe that on Figure 7.6, textual description change along the <`SG2, SR2, FA2, HW1`> trace does not propagate into the functional architecture due to the safety requirement mapping nature of the (`SR2, FA2`) link. On the other hand, SIL change does propagate, as `FA2` got connected to a new safety requirement; thus, its SIL has to be potentially (and in this case, also actually) modified.

Value-based propagation can further reduce the extent of the change impact cover. In addition to types, it also takes into account the nature of the propagating changes as well as the current values captured in each requirement. Notice that HW1 has not to be changed, although `FA2` has been mapped to it. The reason is that although it got newly connected to a (newly) high-SIL function, it already has the highest SIL level. Thus, during marking, propagation can safely stop here. Figure 7.7 demonstrates the relationship between the change impact cover extents and the resolution of propagation.

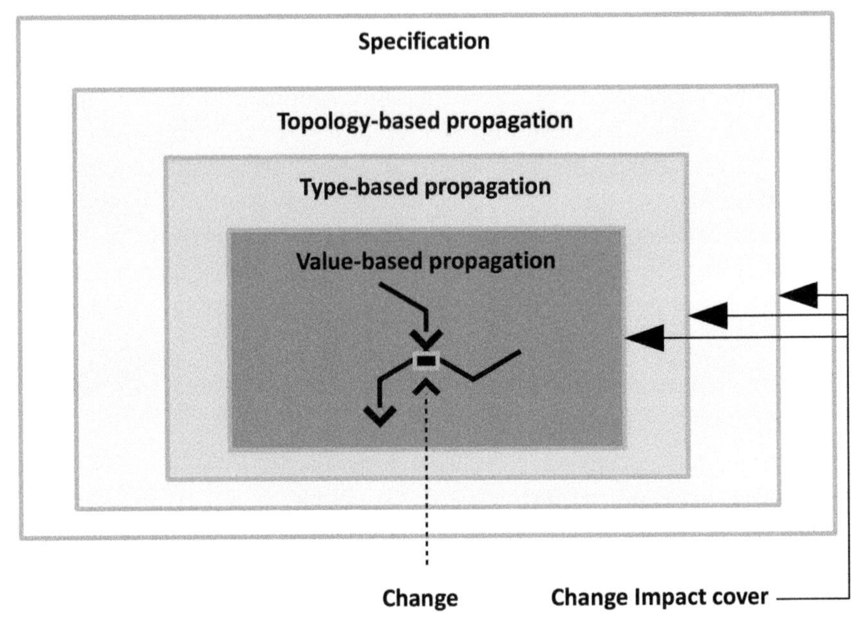

Figure 7.7 Propagation resolution and computed change impact cover extent.

Independently of the category, we have to note that propagation does not necessarily happen only "forward" or "downstream". The change of a requirement may impact its parent (containment-wise), not just its children; and it may impact the sources of its incoming traceability links, not just the targets of its outgoing ones. Tooling supports the user to configure the directionality of propagation; this is a largely orthogonal concern to the propagation resolution. For the sake of simplicity, in the following, we focus on the forward direction unless otherwise noted.

7.5 Abstraction Levels of Impact Propagation

We have argued informally that there are three major categories of change impact propagation from resolution. In this section, we describe and compare these categories using a simple example.

Let us consider the rich requirement structure in Figure 7.8. In addition to an SIL attribute, our requirements can have a priority attribute, too. In the context of this example, priority expresses the importance of overall operational efficiency with the levels `HIGH,` `MEDIUM` and `LOW.` It aggregates

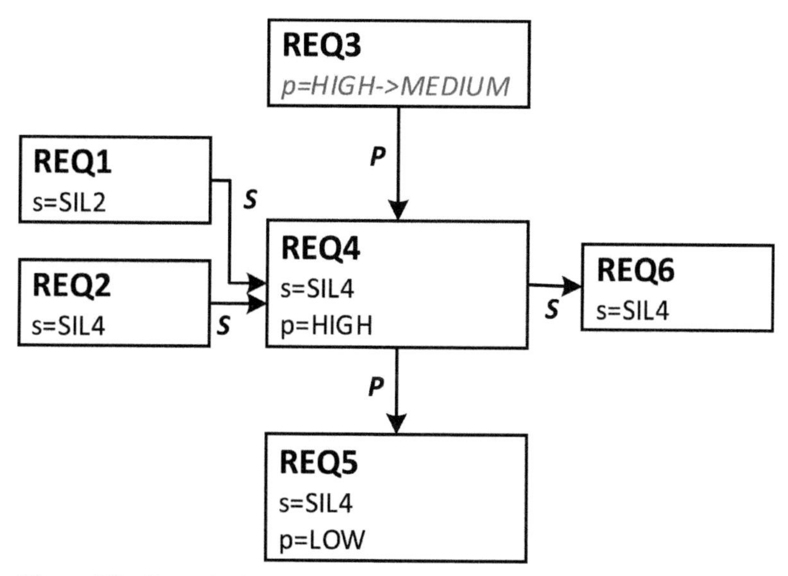

Figure 7.8 Example rich requirement structure for propagation categorization.

some concepts, including maintainability, time to repair and cost of operation. This priority concept is largely independent of SIL, and while the safety level is an absolute requirement, priorities are subject to business considerations and not critical to meet.

The example requirement model uses

- containment (parent relationship, denoted by P),
- SIL-mapping traceability links (denoted by S) and
- SIL and priority level attributes for the requirements.

For the purposes our example, we assume that the following consistency rules are applied during requirement management.

- **Rule 1**. Any requirement that has an incoming "safety mapping" (S) traceability link has an SIL attribute, and its value is the maximum of the SIL values at the source requirements. For codification, the requirement engineer should be able to derive this rule from the process definition and the safety standards that have to be applied.
- **Rule 2**. A prioritized requirement must have only prioritized descendants. This value can be only less or equal than that of the parent. For codification, the requirement engineer should be able to formulate this rule as a locally used and observed rule.

Let us emphasize that these rules are for demonstration purposes. SIL value constraints along traceability links can be much more complicated in the general case. The handling of priorities also represents only one possible choice; among others, even its exact reverse may be justified in a specific project. Our modeling approach and the subsequently introduced solution method can support almost arbitrary rules. We also handle the potential non-determinism of the rules.

The change we will be concerned with is modifying the priority of `REQ3` from `HIGH` to `MEDIUM`. Figure 7.9 demonstrates propagation for the three categories.

7.5.1 Topology-Based Propagation

We can propagate the change along the outgoing dependencies (containment and traceability links) of the requirement, marking requirements transitively as *SUSPICIOUS*. This approach is commonly referred to as *topology-based* change propagation. In addition to attribute changes, the creation of new dependencies as well as deletion of existing ones (through deleting the source/target requirement or otherwise) is seamlessly supported.

7.5.2 Type-Based Propagation

The next level is *type-based* propagation. Dependencies have types, as well as the attribute of the originally changed requirement that is changed. We can filter propagation for dependency type as well as changed attribute type. We reflect the changes that are allowed to propagate into the first-level dependents by marking the dependents or some of their own attributes as *SUSPICIOUS*. We can then perform propagation from this first level transitively by propagating the requirement or attribute *SUSPICIOUS* marking using the same configurable filtering and configurable attribute marking mapping mechanism. In the context of the example of Figure 7.8: for priority changes, we propagate the *SUSPICIOUS* marking only along the descendants of the changed requirement. Priority attribute markings are mapped into priority attribute markings, as there is no logical dependency between the two attributes in this case. Note that in addition to the orthogonal analysis of attributes, this as well as the next category supports conjoint analysis – we can express when the change of an attribute propagates to another.

Dependency changes are handled similarly to topology-based propagation.

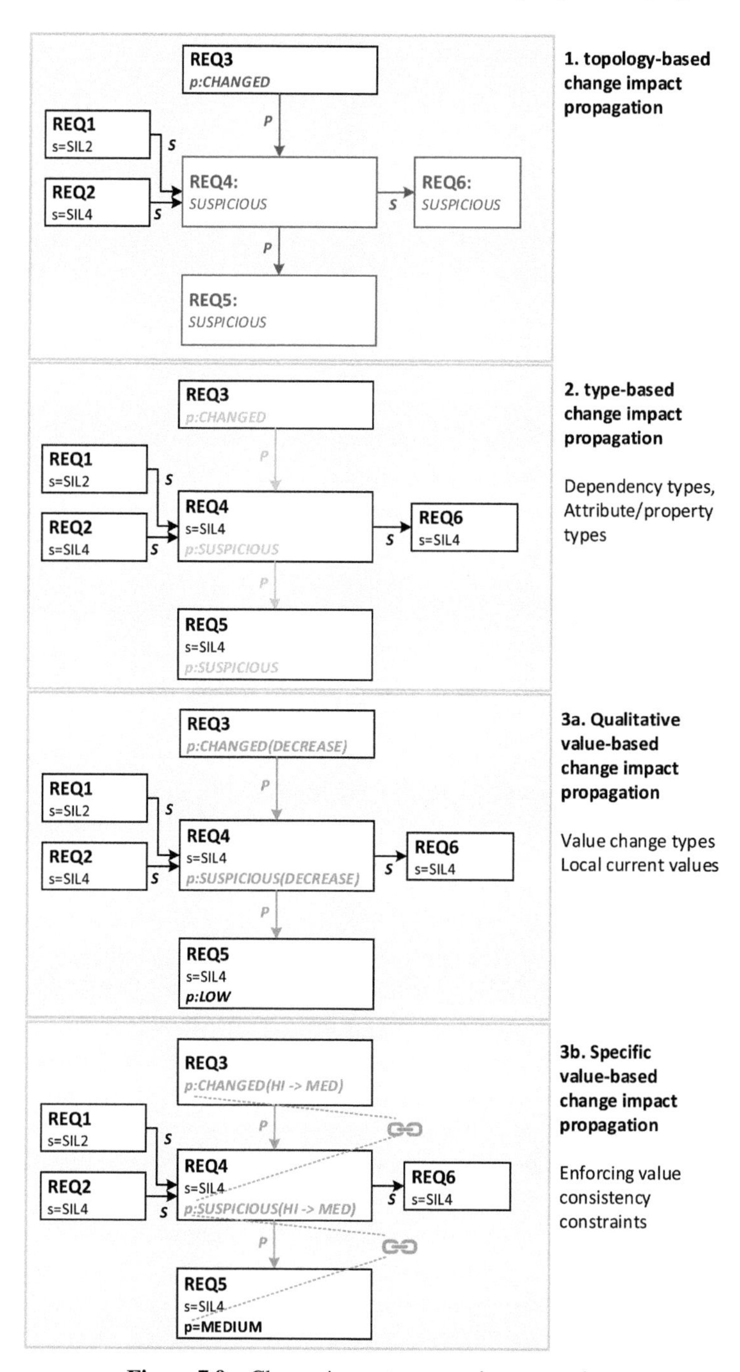

Figure 7.9 Change impact propagation categories.

7.5.3 Value-Based Propagation

Type-based propagation is a powerful tool for controlling the extent of change propagation in specifications that have a dense dependency structure; however, it still does not take into account the *kind* of the change in the value domain. Change can be described either qualitatively or in specific terms; a qualitative description for priorities would be e.g. declaring its *increase* or *decrease*, while the specific description is either the new value or the old-new value pair. We establish the category of *value-based* change impact propagation with these two subcategories.

In the qualitative case, observe that if Rule 2 is known, it means that the priority *increase* change of a parent does not *need SUSPICIOUS*-marking on the children. However, when it *decreases*, priorities in the children *may* have to be revisited. We can say even more: when the priority at the next level is `HIGH`, we are certain that changes are necessary; conversely when it's `LOW`, we are certain that propagation stops here. A somewhat similar qualitative logic exists for the SIL mapping in our case; at level 4, any increase upstream will not have any impact, while a decrease may require requirement reconsideration.

The downside is that the rule set that is to be used has to be defined; however, with predefined templates, this promises to be a manageable overhead.

The next logical step is to consider propagating the specific change and computing the specific local changes that may be necessary to be made. We call this approach *specific value based change propagation*. On paper, this idea seems not too far-fetched (see the example in Figure 7.9). However, it requires formulating explicit, value-specific consistency rules. More specifically, propagation needs value *change* consistency rules that connect allowed changes in localized requirement contexts (in the simplest, the allowed attribute co-changes at the two ends of a typed link). However, it is easy to see that the most of such change consistency rules can be transformed into value-specific specification consistency rules and vice versa. In this last case, the line between propagation and marking processing becomes very blurred, as essentially specific change candidates are computed as a marking during propagation.

We treat this last category as a theoretically interesting option; however, it is one that has little immediate value to the practice in an industry that doesn't even use type-based change propagation in a widespread way yet.

We have conducted an analysis of a sample of the best-of-breed requirement management solutions, to determine the extent and sophistication to which they support assessing the propagation of requirement change impacts. Topology-based propagation seems becoming available. Type-based propagation is still a novel feature, available, e.g., in Rational DOORS Next Generation. Value based propagation (qualitative or otherwise) is practically non-existent yet.

7.6 Resolution Modeling with CSP

To establish a common, computable framework for the first three categories above, we define them declaratively as finite-domain Constraint Satisfaction Problems (CSPs) [9]. The motivation is that many sensitivity analysis tasks in error propagation assessment and test generation are known to be definable and also solvable this way – and tracking the propagating effect of requirement changes is very similar to tracking the potential effects of faults in a system.

In CSPs, a finite set of variables, each with a nonempty domain, is subjected to a set of constraints. Each constraint is a relation that specifies the permissible value combinations for a subset of the variables; a solution of the CSP is such a value-assignment of the variables that satisfies each constraint. An important category of such problems is finite-domain CSP, or csp(FD); in this case, the variables are discrete and have finite domains. This way, csp(FD) expresses combinatorial search style problems.

The power of csp(FD) is that it can be used to declaratively specify a problem and letting one of the mature, optimized and very sophisticated existing tools to look for a solution (or enumerate all solutions). Tools widely recognize a standard set of composable "simple" constraints (linear arithmetic equalities and inequalities, Boolean arithmetic, etc. over the declared variables) and so-called global constraints, too. The latter involve a potentially large number of variables and need specific algorithmic optimization (for an exhaustive list, see the Global Constraint Catalog [10]). Constraint problems have a widely recognized standard representational language in the form of XCSP3 [11].

Change impact propagation problems can be easily represented as a CSP. We declare the marking of each requirement, attribute and dependency (containment and traceability links) as a variable; define the possible marking value set for each; describe propagation as constraints and lastly introduce the constraints for the performed changes. Table 7.1 gives an outline of this process.

Table 7.1 Comparison of change impact propagation categories

Cat	Requirement/ Dependency Marking Literals	Constraints for the Dependencies: If Dependency *d* with Type *t* is …	Constraints for the Dependencies: … then	Further Constraints
1.	Reqs: *CHANGED, SUSPICIOUS, INTACT* Attributes: N/A Dependencies: *DELETED, ADDED, SUSPICION_LINK, INTACT*	Not *DELETED* or *ADDED* the source of *d* is not *INTACT* Not *INTACT*	Target of *d* not *INTACT* *d* is *SUSPICION_LINK* Target of *d* is not *INTACT*	Only the actual changes can be *CHANGED, DELETED, ADDED* Maximize the number of *INTACT* markings
2.	Reqs: N/A Attributes: *CHANGED, INTACT, SUSPICIOUS* Dependencies: *DELETED, ADDED, SUSPICION_LINK, INTACT*	not *DELETED* or *ADDED* attribute *a* at the source not *INTACT* *a* declared propagation source for *t* *DELETED* or *ADDED*	Attributes declared as *a*-propagation target for *t* at the target of *d* not *INTACT* *d* is *SUSPICION_LINK* Attributes at the target declared for *t* creation or deletion propagation not *INTACT*	Only the actual changes can be *CHANGED, DELETED, ADDED* Maximize the number of *INTACT* markings
3a.	Reqs: N/A	Rule set that for each dependency type *t*, encodes the relation expressing the together permissible (or ruled out)		Only the actual changes can be *CHANGED, DELETED, ADDED*

Table 7.1 Continued

For each attribute: *NOCHANGE, SUSPICIOUS_CHTYPE1, SUSPICIOUS_CHTYPE2, ...*	• source-side attribute marking values, • target-side attribute marking values, • dependency markings.
Dependencies: *DELETED, ADDED, SUSPICION_LINK, INTACT*	Rules allowed also to incorporate current attribute values.

A few things have to be noted on this framework. For topology- and type-based propagation, the table describes only rules for forward propagation; however, backward propagation rules can be introduced similarly. Notice that type-based propagation introduces attribute marking and discards requirement marking; the latter can be incorporated (with some complexity increase) or emulated by declaring all attributes suspicious.

For qualitative value-based propagation, due to the variability of the rules (that is the intended goal), we can't characterize propagation rules in the same manner. Still, all practically important propagation intentions can be formulated using standard CSP expressions. For instance, propagation of priority increase suspicion for requirement `R1` and `R2` interconnected through a link type `t` can be expressed e.g. as `t_connected` (**`R1_p_marking`**, **`R2_p_marking`**) `AND` **`R1_p_marking`** == *`SUSPICIOUS_INCREASE`* `AND` **`R2_p_current`** < *`HIGH`* → **`R2_p_marking`** == *`SUSPICIOUS_INCREASE`*.[3] That said, our ongoing work addresses creating a domain-specific language that simplifies the creation of the sets of rules.

7.7 Conclusions

The OMG Requirements Interchange Format (ReqIF) assures interoperability between cooperating partners. This standard is a natural candidate to information exchange between designer and independent software/system verification and validation (ISVV).

Using ReqIF facilitates (which is occasionally only the question of asking the developers using top-end requirement design tools for providing

[3]Variables are typeset bold, value-literals are typeset italic and `t_connected` is a constraint that we defined based on the specification.

the native ReqIF model in addition to the derived documentation not containing the model) an immediate entry to the benefits of lightweight formal models.

At the same time, OMG ReqIF provides a well-regulated set of rules for the developer-ISVV interoperation and the communication of the assessment results.

Traceability is a priority concept in ReqIF. The option of defining the structure the document, introducing types and well-formedness constraints are all major means to introduce the main concepts of domain-specific MBSE.

Similarly to development, where advanced tools can generate traditional office-like documentation out of their internal ReqIFmodels, ISVV can highly benefit from using RequIF as the core model for communicating ISVV results.

Ontologies provide an easy way to overcome the limitations of ReqIF regarding conceptual modeling. Ontology-based metamodel design is a modern model development paradigm, as its standardized language and development tools implement all the main concepts of complexity management, like the composition of complex ontologies out of simpler ones, hierarchical modeling and aspect weaving. At the same time, their well-defined semantics allows using reasoners.

The simple mathematical background of ontologies, set theory results in a low entry threshold related to skills. The built-in logic reasoners can check the contradiction freedom of a requirement set (by a satisfiability test), and its well-formedness (by a subsumption check), thus deliberating the ISVV of tedious manual checks.

Ontologies are highly standardized. Model formats assure interoperability; moreover, standard transformations exist to the world of metamodeling. As ontologies provide an abstract representation of knowledge, automated export and import tools exist between ontologies and knowledge storage tools like structured semi-formal representations (Excel), relational, object-oriented and graph databases.

Classically, requirement changes involve a significant effort and quality cost, especially if the tooling provides no proper guidance for the reassessment. Intelligent change impact analysis helps properly focusing the assessment after a change by evaluating the propagating effects of the introduced changes. In a properly structured requirement specification with a rich traceability structure, algorithmic analysis can significantly reduce the extent of the change impact propagation cover that analysts have to check.

References

[1] Object Management Group. (2017). *Requirement Interchange Format (ReqIF)*. Available at: http://www.omg.org/spec/ReqIF/ (accessed on 1 March 2017).

[2] *Requirements Management for Eclipse*. Available at: https://eclipse.org/rmf/ (accessed on 1 March 2017).

[3] Eclipse. (2017). *ProR Requirements Engineering Platform*. Available at: http://www.eclipse.org/rmf/pror/ (accessed on 1 March 2017).

[4] Knublauch, H. (2004). "Ontology-driven software development in the context of the semantic web: An example scenario with Protege/OWL," in *1st International Workshop on the Model-Driven Semantic Web (MDSW2004)* (New York, NY: IEEE), pp. 381–401.

[5] W3C. (2009). *W3C: OWL 2 Web Ontology Language Document Overview*. Available at: https://www.w3.org/2009/pdf/REC-owl2-overview-20091027.pdf (accessed on 1 March 2017).

[6] ISO. (2007). *ISO/IEC 24707:2007: Information technology – Common Logic (CL): a framework for a family of logic-based languages*.

[7] Pataricza, A., Gönczy, L., Kövi, A., and Szatmári Z. (2011). "A Methodology for Stand-ards-Driven Metamodel Fusion," in *Model and Data Engineering: First International Conference, MEDI 2011* (Berlin: Springer), 270–277, Óbidos, Portugal, September 28–30, 2011. Eds L. Bel-latreche and F. Mota Pinto.

[8] Tarnai, G., and Sághi, B. (2006). "Hazard and Risk Analysis of Human-Machine Interfaces of Railway Interlocking Systems," in *7th World Congress on Railway Research*, Montral, Canada, 4–8 June.

[9] Brailsford, S. C., Potts, C. N., and Smith, B. M. (1999). Constraint satisfaction problems: algorithms and applications. *Eur. J. Operat. Res.* 119.3, 557–581.

[10] Beldiceanu, N., Carlsson, M., and Rampon, J.-X. (2012). "*Global Constraint Catalog, (revision a)*." Available at: http://www.diva-portal.org/smash/record.jsf?pid=diva2:1043063 (accessed on 1 March 2017).

[11] Frédéric, B., Lecoutre, C., and Piette, C. (2016). "*XCSP3 Specifications – Version 3.0*." Available at: http://www.xcsp.org (accessed on 1 March 2017).

[12] RODIN. (2017). *Rigorous Open Development Environment for Complex Systems*. Available at: http://rodin.cs.ncl.ac.uk/ (accessed on 1 March 2017)

[13] Object Management Group. (2017). *Ontology Definition Metamodel (ODM)*. Available at: http://www.omg.org/spec/ODM/ (accessed on 1 March 2017).

[14] Government of the United Kingdom, Department of Transport. (2017). *Rail Accidents and Safety Statistics Tables*. Available at: https://www.gov.uk/government/statistical-data-sets/rai05-rail-accidents-and-safety (accessed on 1 March 2017).

8

STECA – Security Threats, Effects and Criticality Analysis: Definition and Application to Smart Grids

Mario Rui Baptista[1], Nuno Silva[1], Nicola Nostro[2], Tommaso Zoppi[3,4] and Andrea Ceccarelli[3,4]

[1]CRITICAL Software S.A., Coimbra, Portugal
[2]Resiltech s.r.l., Pontedera (PI), Italy
[3]Department of Mathematics and Informatics, University of Florence, Florence, Italy
[4]CINI-Consorzio Interuniversitario Nazionale per l'Informatica-University of Florence, Florence, Italy

8.1 Introduction

The reliability of electrical power systems, since their first use, has been addressed focusing on ensuring the continuous power supply and on the management of critical situations in order to avoid electrical disruption due to potential failures. In the last decade, we are witnessing the increasing development of Smart Grids, with €3.15 billion investment in Smart Grids projects amongst the EU-28 Member States only in the period 2002–2014 [1]. Smart Grids enhance the classical electrical systems by introducing optimization of grid management, both from transmission and quick reaction to power disruption through real-time and automated technologies; deploying and integrating of large-scale renewable energy systems; reducing management and power costs, for final users; and introducing and integrating of smart appliances and consumer devices. While these new aspects make the electrical systems effective, they become more and more interconnected thus making them vulnerable to cyber and physical attacks [2–4]. Indeed, it is possible to remotely perform changes (e.g., to instructions, commands and configurations), disabling actions, shut down or in general interfere with the proper functioning of the system, thus causing in the worst case significant damages and safety issues [3, 5].

In the Smart Grid domain, security threats can be originated by several agents: consumers, insiders, and terrorists [3]. Customers could be interested in falsifying smart meters data in order to steal electrical power. Similarly to attacks performed to broadband modems, customers may try to attempts attacks to smart meters aiming at modifying the firmware controlling the reporting operation, thus decreasing the usage of electricity [3]. Terrorist attacks to smart grids may lead to unprecedented black-outs, from the point of view of spatial and time extension [4]. This calls for a fundamental attention to the identification and management of potential security threats.

This chapter proposes the STECA (STECA – Security Threats, Effects and Criticality Analysis) approach to perform security assessment of Smart Grids. The hereby proposed process describes a way in which to identify vulnerabilities, their related threats, and proposes a risk assessment approach and a path to identify appropriate countermeasures. This process is based on the same principles used for the Failure Mode and Effect Analysis (FMEA)/FMECA process, which is a technique widely used for safety critical analysis and is highly regarded by the majority of international standards [6]. STECA starts from a vulnerability point of view and moves on towards threat analysis and criticality assessment. Following the guidelines defined in [7], the approach is instantiated on a Smart Grid use case, resulting in a set of precise guidelines and a systematic way to perform security assessment including vulnerability evaluation and attack impact analysis.

8.2 Motivation

8.2.1 Motivating Concerns in Industry

A fundamental aspect that has to be considered in the implementation of Smart Grids and that is currently under the stakeholders' spotlight is related to the security issues yet to unveil in the overall Smart Grid or at the connected devices [3, 8], and the consequent impact on safety. Among the impact situations of a service disruption due to a cyber or physical attack, *property/financial damage* and *human life hazard* should be kept in closer consideration, as the time for recovering is currently unpredictable.

Previous works on Smart Meters qualification revealed potential security weakness and exposed some of the equipment vulnerabilities [3]. This can present a great risk for the future implementation of Smart Grids. It is also true that, due to the ground-breaking character of this technology and the quantity of interfaces that are made available, the *security requirements of the components that will operate in the grid and the grid system itself are not yet sufficiently accurate* (either they are not studied or implemented/tested, not analysed or imposed yet at a larger scale). This is also strengthened if we consider these systems general exposure in terms of pervasive interfaces.

In fact, in an informal *industrial security assessment of Smart Grid* components, the company CRITICAL Software identified security problems that are usually disregarded by traditional assessment approaches if performed without a proper process or tools. Examples of these problems included: (i) denial of service possibility, (ii) proscribed access to equipment; (iii) physical security deficiencies; (iv) unintended access to systems parameters that should be read-only; and (v) communication protocol implementation and specification weakness. The experience of CRITICAL Software's industrial assessment projects ended up providing most of the incentive for the development of the STECA process due to the gaps found. First, CRITICAL Software was providing a security assessment to substantiate a test framework being developed at the time. Security issues were observed in the assessed Smart Grid components, both on requirement analysis and actual component functional tests, despite the work was focused only on a limited part of the target Smart Grid equipment. In yet another case only the communications protocols were under test on a preliminary stage. For instance, it was identified that it is possible without much trouble to generate conditions that force the interruption of energy supply to a user on the grid. Either by simulating excessive energy demand or by tampering with billing contract parameters, it is also possible to provoke a Denial of Service. This form of service disruption by hacking the metering equipment is a commonly acknowledged threat, but the impact is largely underestimated. Several other ways of generating conditions that will switch off the Load Control Switch can also be identified. It was also clear from the functional testing that the meter's communication ports could easily be disabled by setting its timeout parameter to zero, rendering the equipment incommunicable and thus impossible to be reconfigured remotely.

Though this experience identified serious security impact scenarios that justified the need of security assessment, the support available today to security assessment is limited. There is no history on the components usage and thus no clear way to understand the attack trends or attacker profiles, the attacker objectives and the effects of the attacks. It is extremely difficult to rate the likelihood of a threat, on which to perform a risk or hazard assessment. Summarizing, there are no real data to work on, which obstacles the possibility to create a solid base to build a security assessment upon. Also, as there is no relevant history of these analyses, it is impossible to even start by using previous knowledge, checklists, pre-defined lists, etc.

One should also consider the constant struggle that resides in identifying the vulnerabilities and security threats. On a system of this sort the number and diversity of security threats could be huge. An undertaking of this magnitude should inevitably find trouble when aiming to achieve completeness: to claim that all vulnerabilities have been identified and all security threats analysed will prove to be a nearly impossible task. Even an expert experience based approach to identify a procedure to tackle this problem is not a straightforward exercise.

8.2.2 State of the Art and Background

Standards such as [9–11] propose general, high level methodologies to guide the security assessment of systems. However, standards typically present the main steps but they do not describe the techniques that can be exploited to realize these steps. This calls for solutions that, still maintaining compatibility with the standards, are able to provide an adequate support to the security engineers. Additionally, several challenges are still open, such as verifying the completeness of an analysis or compute likelihood and impact of a given threat.

Several works target techniques for security assessment, also considering interdependencies between security and safety. The work presented in [12] proposes an extension of the FMEA safety analysis technique, aiming to analyse likelihood and impact of cyber-attacks to embedded multicore systems in the automotive industry. Another contribution, still related to the automotive domain [13] aims at proposing a novel approach to deal with both safety hazard and security threat analysis combining the Hazard Analysis and Risk Assessment with the security STRIDE approach for the automotive battery management [14] proposes a framework for quantitative security analysis used to identify potential attack points and paths, thus to recognize those that are feasible from the perspective of an attacker and finally proposing meaningful countermeasures to the system. In [7] the authors propose a general methodology to understand issues' criticality and the difficulties in finding a proper solution able to deal with interdependencies between safety and security. To this purpose, in their work a general security threats library has been developed, which can be updated over the time and has been mapped to the NIST security controls [8]. Other contributions evaluating the effects of security breaches exploits exist as the work in [15], which states a comprehensive and practical framework for electric smart grid cyber-attack impact analysis using graph-theoretic dynamical systems paradigm.

The STECA process presented in this chapter is specifically focused on Smart Grids. It naturally includes the objective of detecting potential security threats and providing efficient mitigations, and it translates the concept of FMEA to a *vulnerability-oriented* security assessment where reference categories are extracted from supporting *threat libraries*. Additionally, it guarantees compatibility with main standards [9–11]: in fact, the reference data to build the threat libraries are extracted from the standard [10], and it is easy to define a correspondence between the main steps of the STECA process and the steps of methodologies in [9, 11].

8.3 STECA Process Description

This section presents a detailed description of the STECA process, along with a running example to illustrate the application of the process to an actual industry problem related to the main theme of this publication.

8.3.1 The High Level STECA

The hereby proposed process (Figure 8.1) describes a way to identify vulnerabilities, their related threats, proposing both a risk assessment approach

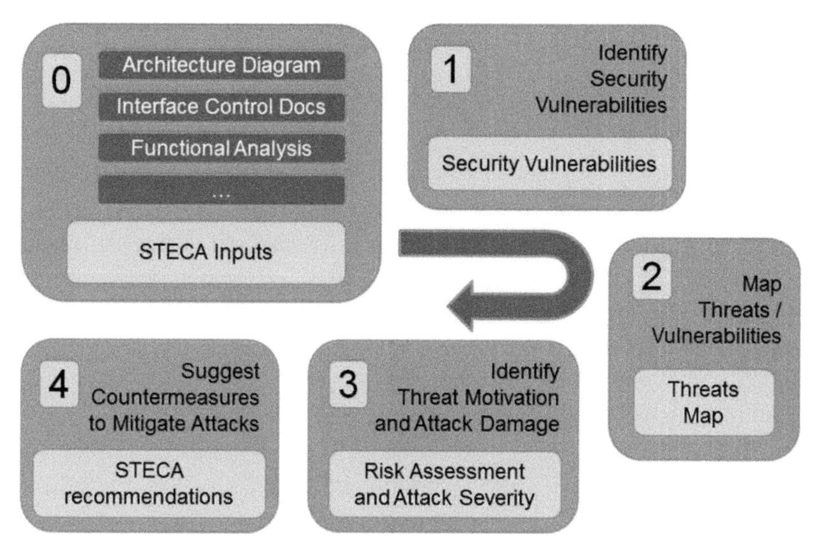

Figure 8.1 High level view of the STEC process.

and a path to identify appropriate countermeasures. Four high-level steps are identified.

This process is based on the same principles used for the *FMEA/FMECA* process [12] which is widely used for safety critical analysis, and is highly regarded by the majority of international standards. Subsequently, the high level steps depicted in Figure 8.1 will be described in closer detail.

8.3.2 STECA Inputs

In order to efficiently apply the process several inputs are required and need to be collected. The input set includes, but is not limited to:

1. *The Architecture Diagrams*. These, along with the Functional Analysis, will be used to identify the system's vulnerabilities.
2. *The Interface Control Documents*. These will allow a better threat identification while analysing vulnerabilities.
3. *A Functional Analysis*. This, along with the Architecture Diagrams, will be used to identify the system's vulnerabilities.
4. *Other useful input information*. Typical security attacks, history data, system requirements, environment conditions, requirements, etc.

For the running example we're using the diagram in Figure 8.2 – an energy industry Smart Grid, focusing on the Smart Meter Home Area Network (HAN) – the most widespread case of user connected to the Smart Grid, also a high vulnerability spot as it exposes the metering devices to the internet through the Consumer HAN.

8.3.3 Security Vulnerabilities

With the STECA inputs we can identify possible intrusion and attack locations considering the system weak spots listed in Table 8.1. For each of them, we reported an extended description and the links to the consolidated ISO/IEC 27005 [6] vulnerability classification which lists the hardware, network and software vulnerability categories. Additional vulnerability classifications are the Microsoft Security envelopment Lifecycle (SDL) [16] and the CWE3 (Common Weakness Enumeration [17]), which is a detailed and community-developed list of common software weaknesses.

Traditionally, the vulnerability assessment [2, 11] of architectures such as the one in Figure 8.2 are performed by (i) cataloguing assets and capabilities (resources) in a system, (ii) assigning quantifiable value (or at least rank order) and importance to those resources, (iii) identifying the vulnerabilities

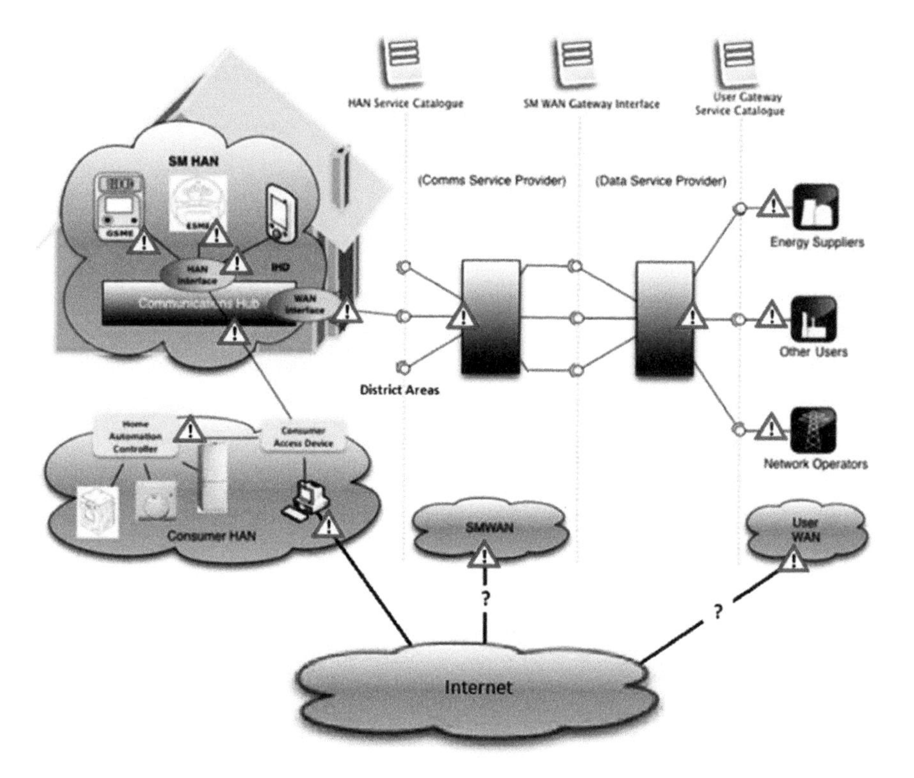

Figure 8.2 Example from the Energy industry showing the architecture of a Smart Grid.

or potential threats to each resource and, (iv) mitigating or eliminating the most dangerous vulnerabilities for the most valuable resources.

The first three steps are required to be performed in order to obtain a vulnerability list. Also, by assigning an order (value) to the resource (vulnerability) we are simplifying the threat severity definition described in Section 8.3.5.

Each component (system resource) should be classified with a value (as of an asset) which could simply be a traditional High, Medium or Low grade according to the associated monetary replacement cost – to be defined by the system/subsystem owner; and a severity grade based on the impact that its failure would inflict on the system. To do this, the catalogue depicted later in Section 8.3.5 is proposed to be used.

Continuing the running example, and focusing on the Communications Hub (in the Smart Meter HAN) as it is a gateway to the metering devices, and

Table 8.1 Vulnerabilities, weak spots, and security threats

Vulnerabilities	Weak Spots	Weak Spots	Security Threats
Network	CH communications protocol Smart Meter access control	CH communications protocol	Message Modification Man in the middle Footprinting
Software	CH communications protocol Smart Meter access control Smart Meter functions	Smart Meter access control	Unauthorized access Password cracking Disclosure of confidential data
Hardware	Smart Meter functions	Smart Meter functions	Conduct cyber-physical attacks on organizational facilities Arbitrary code execution

the electricity Smart Meter itself, as it is a big concern in the motivation, we obtain the following:

- **Communication Hub**: *Value*: Low; *Severity:* Minor;
- **Electricity Smart Meter**: *Value* Low; *Severity:* Critical/Catastrophic.

8.3.4 Threats Map

In this step of the process the security threats shall be identified and catalogued by performing the following sub-steps:

- *Identify the threats for each vulnerability*. Following the list produced in the previous step, we list the threats that may exploit each vulnerability. This operation will produce a list of threats per vulnerability.
- *Catalogue Threats (NIST classes)*. Identified threats will most likely be found in the known threats list, thus having associated countermeasures. Most possibly, the gathered threats have already been identified in different contexts and catalogued in a generic fashion in existing documents, thus a set of countermeasures and preventive actions might already be available. For this purpose, already existing classification taxonomy may be used. For this process we're using the threat library already created in [7], which will help to catalogue the threats to the NIST classes and the suggested countermeasures.

- *Threat Classification Completeness Check.* If unlisted threats arise, countermeasures should be suggested to mitigate them and the threat library should be complemented by adding this new information to the respective taxonomy class.

Next in the running example, the weak spots are identified and respective Vulnerability categories. Some examples are reported in Table 8.2.

Following through with the running example and mapping these threats to the Threat Library it is possible to catalogue almost all of them to the NIST classes and gather the respective countermeasures to mitigate them. The NIST 7628 [9] states, in more than one occasion, that its focus is cyber-security and therefore "The requirements related to emergency lighting, fire protection, temperature and humidity controls, water damage, power equipment and power cabling, and lockout/tag-out are important requirements for safety. These are outside the scope of cyber security and are not included in this report. However, these requirements must be addressed by each organization

Table 8.2 Linking weak spots and ISO/IEC 27005 vulnerability categories

Weak Spots	Description	ISO/IEC 27005 [6] Categories	Threat Example
Component interfaces, communica-tions ports/ protocols	These are usually the targets to corrupt communications either to attain disruption or impersonating another party.	Network	Man in the middle packet sniffing conducted between the smart meter and the Energy Management Gateway
		Software	Inject malicious code into the USB device controller (*BadUSB*, [18])
Memory and Storage Units	These may be used to store malicious code for later execution or even altered firmware when system reconfiguration is required.	Software	Installation of a malware which damages user data or key memory areas
		Hardware	Damaging the Hard Drives (i.e., putting a magnetic source near the storage rack servers)
Processing Units	These, of course, may be used to execute the malicious code.	Software	Inserting malicious code that calls for ALU operations slow downing the whole execution
		Hardware	Malicious hardware module targeting the performance of the cache accesses or generating power faults

in accordance with local, state, federal, and organizational regulations, policies, and procedures." In this example, the "Conduct cyber-physical attacks on organizational facilities" threat could be the example stated in the motivation section where an Electricity Smart Meter is rendered inoperative and incommunicable, inducing a denial of service occurrence with potentially catastrophic impact, and will be the focus of the running example in the following sections.

8.3.5 Risk Assessment and Attack Severity

For this step, similar to the *Cause and Effects analysis*, there are two things that need to be accounted for when considering each threat event: *probability* and *impact*.

Probability: (Attack Profit – Motivation). In several contexts the parameters used to calculate the probability of an attack are based on a likelihood extracted from existing data. In general, there are no "reasonable" approaches to compute the likelihood of an attack, apart using past history, meaning that this specific approach is applicable only for few systems. As in this case there is no such data, we propose to use the estimated benefit that the attacker may obtain due to a successful attack. This can be seen as a combination of (i) *Cost:* availability of resources to perform the attack (time, money, state of the art hardware), (ii) *Risk of detection* (to what extent can the attacker hide his actions and how much does he care about being detected) and (iii) *Payoff* (the benefit that an attacker expects from exploiting a vulnerability). These three components can be considered separately or grouped together to build a unique likelihood score that can be obtained depending on the specific needs. One possible likelihood example could be an index that represents the cost/benefit trade-off, calculated as the fraction of *Payoff* over *Cost* but, for this purpose, we propose a form of calculation using the three variables as shown in Figure 8.3.

Having the lowest values on the origin (0,0,0) and increasing each variable in each of the respective axis. Colour code (green, yellow and red) represents the likelihood of an attack as depicted (unlikely, moderate and likely, respectively).

The Attack Probability assigned values are just an example and, when applied, should be adapted to the respective domain requirements. If it is considered that a system exposure to attack is less dependent on payoff than the other variables, more red and yellow dots should be reflected on the graph.

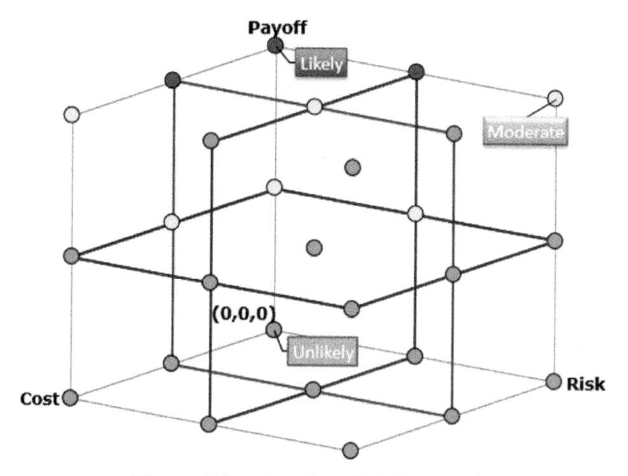

Figure 8.3 Attack probability graph.

Impact: (Attack Damage). The extent to which an attack may cause damage. This should include all harmful consequences. It may be calculated based on the individual resources involved (associated value and severity), the effects produced by a general failure of the resources involved and the derivate pernicious effects from the aftermath. The worst case scenario should be considered for the Impact calculation.

The Risk of a given Threat Event used in this case is based on a traditional Criticality approach. The values used in this example have a higher weight on the Impact rather than the Probability.

The values in Figure 8.4 were calculated by multiplying the grades assigned to the respective probability and impact ranks. A green code was assigned to values lesser than or equal to 3, yellow to values between 4 and 9 inclusive and red to values greater than or equal to 10. Again, this is an illustrative example and the colour code should be adapted to the

Threat Event Risk		Attack Probability		
		Likely (3)	Moderate (2)	Unlikely (1)
Impact Severity	Catastrophic (5)	15	10	5
	Critical (4)	12	8	4
	Major (3)	9	6	3
	Minor (2)	6	4	2
	Negligible (1)	3	2	1

Figure 8.4 Threat Event Risk Matrix.

	Severity	Description of Consequences
Category	Catastrophic	The threat event could be expected to have **multiple severe or catastrophic** adverse effects on organizational operations, organizational assets, individuals, other organizations, or the Nation.
	Critical	The threat event could be expected to have a **severe or catastrophic** adverse effect on organizational operations, organizational assets, individuals, other organizations, or the Nation. A severe or catastrophic adverse effect means that, for example, the threat event might: (i) cause a severe degradation in or loss of mission capability to an extent and duration that the organization is not able to perform one or more of its primary functions; (ii) result in major damage to organizational assets; (iii) result in major financial loss; or (iv) result in severe or catastrophic harm to individuals involving loss of life or serious life-threatening injuries.
	Major	The threat event could be expected to have a serious adverse effect on organizational operations, organizational assets, individuals other organizations, or the Nation. A serious adverse effect means that, for example, the threat event might: (i) cause a significant degradation in mission capability to an extent and duration that the organization is able to perform its primary functions, but the effectiveness of the functions is significantly reduced; (ii) result in significant damage to organizational assets; (iii) result in significant financial loss; or (iv) result in significant harm to individuals that does not involve loss of life or serious life-threatening injuries.
	Minor	The threat event could be expected to have a limited adverse effect on organizational operations, organizational assets, individuals other organizations, or the Nation. A limited adverse effect means that, for example, the threat event might: (i) cause a degradation in mission capability to an extent and duration that the organization is able to perform its primary functions, but the effectiveness of the functions is noticeably reduced; (ii) result in minor damage to organizational assets; (iii) result in minor financial loss; or (iv) result in minor harm to individuals.
	Negligible	The threat event could be expected to have a negligible adverse effect on organizational operations, organizational assets, individuals other organizations, or the Nation.

Figure 8.5 Description of impact categories.

domain requirements. Higher a criticality is inherent the intervals should slide accordingly.

As for the Severity categories, the consequences were gathered from the NIST Threat Events Impact Assessment Scale but, once again, they should be dependent on the domain requirements. Discrimination goes as in Figure 8.5.

Picking up the running example, to assess a likelihood value for the "Conduct cyber-physical attacks ..." threat by using the suggested process, we would come out with the following result: *Cost: low, Risk: Medium, Payoff: Medium/High.*

In the case of the Payoff the assigned grade may depend on the objective of attacker. If the objective is the actual denial of service, the Payoff could be considered High – the worst case scenario. This would produce a Probability result of Moderate to Likely (2 or 3 in Figure 8.5). Moving to the Threat Event Risk calculation, and considering the Smart Meter asset severity grade of Critical/Catastrophic (4 or 5 in Figure 8.5), the result would come out in the range of 8 to 15 (mostly Red).

8.3.6 STECA Recommendations

After all vulnerabilities and respective threats are considered and analysed, countermeasures and preventive actions should be suggested for each of them.

Either from the existing documentation and standards and the educated analysis performed where the vulnerabilities are yet to be acknowledged. Countermeasures should be suggested according to their respective mitigation type, as in:

- *Vulnerability*: the optimal option if possible. If a vulnerability may be avoided all the associated threats will be cleared.
- *Threat Event*: if a treat event may be prevented, the associated security threat will be cleared.
- *Threat probability/impact*: If it is impossible to avoid a threat, consideration should be given to reducing its impact. By downsizing the probability and/or the impact its risk will be downgraded making the system a bit safer. The priority will be set according to the domain and/or system requirements.

The countermeasures are not mutually exclusive and more than one might be applied for each threat. There are, of course, a number of considerations while selecting from the available options, most typically the trade-off between the countermeasure implementation costs vs. its effective security improvement. For a better evaluation in this regard, further iterations of the process including the countermeasures implementations should be performed.

To aid and formalize the process of the security threats analysis, a STECA report depicted in Figure 8.6 should be produced based on the proposed template. For each security threat one of these entries should be included (the fields should be self-explanatory once one is acquainted with the process):

1: STECA ID	4: Veak Spot (Vulnerability)	5: Vulnerability (ISO/IEC 27005 connected categories)	6: Security Threats	7: Threat Library Mapping	8: NIST Proposed Countermeasures	12: Treat Event Risk	13: Alternative Countermeasures	14: Recommendations
STECA-001-01	Smart Meter functions	- Software - Hardware	Execute or generate conditions that lead to the execution of unscheduled functions	Arbitrary code execution	Stay current with patches and updates to ensure that newly discovered buffer overflows are speedily patched.	6		
STECA-001-02	Smart Meter functions	- Software - Hardware	Induce a power cut and reconfigure the communications port to render the meter inoperative and incommunicado	Conduct cyber-physical attacks on organizational facilities	None	10	Use a Smart Meter with redundant metering equipment activated through manual bypass	Should be mandatory for households having infants, seniors or patients on life support systems

Figure 8.6 STECA report example.

1: STECA ID – Unique identifier of a security threat;
2: Architecture Diagram/Model – Relevant Model and/or Diagram files for the process;
3: Domain – Domain to which the process will adapt (Space, Automotive, Railway, Energy...);
4: Weak Spot (Vulnerability) – A mark on the Diagram/Model to signal a weak spot on a component (as in Table 8.1);
5: Vulnerability (ISO/IEC 27005 connected categories);
6: Security Threats – Threat on a vulnerable component (weak spot);
7: Threat Library Mapping – Respective threat in the Threat Library;
8: NIST Proposed Countermeasures – Countermeasure info from the Threat Library;
9: Countermeasure Effectiveness – Applicability of the Threat Library proposed countermeasure to this specific security threat;
10: Attack probability – Calculated as described in Section 8.3.5;
11: Impact Severity – Calculated as described in Section 8.3.5;
12: Treat Event Risk – Calculated as described in Section 8.3.5;
13: Alternative Countermeasures – Countermeasure suggestion if none are available or are considered ineffective;
14: Recommendations – Further considerations to be kept in mind;
15: Assumptions – Assumptions to security threat or regarding information if any;

Notes: Any additional information that might be relevant and does not fit any of the previous.

Note that some of the columns in Figure 8.6 are hidden considering only the most relevant ones for the example. After the report is concluded, meaning all the threats in all the weak spots are analysed and addressed, the STECA process iteration is finished.

To conclude the running example, and as far as countermeasures are concerned (apart from the ones suggested by the Threat Library as shown in Figure 8.6), the suggestions could be something along the lines of the physical countermeasures referred in Section 8.2, filling in the gap in the Threat Library:

- Dumb Meter Bypass
- Smart Meter Black Box

Even if cyber security issues are addressed by threat and risk assessment processes, the STECA can help to identify unaddressed high impact security issues, and support a security/safety report to deliver to the proper authorities.

Based on the STECA results new security requirements may be derived or the existing ones may be improved; those new/updated requirements will lead to improvements in the system safety architecture and design.

8.4 Conclusion

In this chapter, we presented the STECA (Security Threats Effects and Criticality Analysis) process to help formalizing the security analysis of complex systems such as Smart Grids. The necessity of devising STECA stems from the direct experience of engineers working in the security assessment of the Smart Grid domain. The proposed process is established on a similar mature technique used for safety critical systems for decades (FMEA/FMECA) and maps to the well-known NIST taxonomy for the security vulnerabilities and threats analysis. We demonstrated that STECA application is straightforward and useful for security assessment.

References

[1] European Commission. (2014). *JRC Science and Policy Reports, Smart Grids projects outlook.*

[2] European Union Agency for Network and Information Security (ENISA). (2013). Smart Grid Threat Landscape and Good Practice Guide.

[3] Parks, R.C. (2007). *Advanced Metering Infrastructure Security Considerations*. Sandia Report SAND2007-7327.

[4] National Research Council. (2012). *Terrorism and the Electric Power Delivery System.* Washington, DC: The National Academies Press.

[5] NIST. (2011). *NIST Special Publication 800-82, Guide to Industrial Control Systems (ICS) Security.*

[6] International Organization for Standardization (ISO). (2008). *ISO/IEC 27005, Information technology – Security techniques – Information security risk management.*

[7] Nostro, N., Bondavalli, A., and Silva, N. (2014). Adding Security Concerns to Safety Critical Certification," in *Software Reliability Engineering Workshops (ISSREW)* (New York, NY: IEEE), 521–526.

[8] NIST. (2013). *Joint Task Force Transformation Initiative, Security and privacy controls for federal information systems and organizations NIST SP 800-53r4.*

[9] NISTIR. (2014). *NISTIR 7628: Guidelines for smart grid cyber security strategy and requirements.*

[10] NIST. (2013). *NIST Special Publication 800-53 Revision 4: Security and Privacy Controls for Federal Information Systems and Organizations.*

[11] NIST. (2012). *Special Publication 800-30 Revision 1: Guide for Conducting Risk Assessment.*

[12] Schmittner, C., Gruber, T., Puschner, P., Schoitsch, E. (2014) "Security Application of Failure Mode and Effect Analysis (FMEA)," in *Computer Safety, Reliability, and Security*, eds A. Bondavalli, F. Di Giandomenico. SAFECOMP 2014. Lecture Notes in Computer Science, Vol. 8666. Springer, Cham.

[13] Macher, G., Höller, A., Sporer, H., Armengaud E., and Kreiner C. (2015) A Combined Safety-Hazards and Security-Threat Analysis Method for Automotive Systems, in *Computer Safety, Reliability, and Security*, eds Koornneef, F. and van Gulijk, C. Lecture Notes in Computer Science, Vol. 9338. Springer, Cham.

[14] Nostro, N., Matteucci, I., Ceccarelli, A., Di Giandomenico, F., Martinelli, F., and Bondavalli, A. (2014) On Security Countermeasures Ranking through Threat Analysis," in Computer Safety, Reliability, and Security, eds A. Bondavalli, A. Ceccarelli, F. Ortmeier. SAFECOMP 2014. Lecture Notes in Computer Science, Vol. 8696. Springer, Cham.

[15] Kundur, D., et al. (2010). "Towards a framework for cyber attack impact analysis of the electric smart grid," in *Smart Grid Communications (SmartGridComm)* (New York, NY: IEEE).

[16] Microsoft. (2010). *Security Development Lifecycle.*

[17] Common Weakness Enumeration. (2017) *A Community-Developed Dictionary of Software Weakness Types*. Available at: https://cwe.mitre.org/index.html

[18] Kaspersky Lab. (2014). *Release of Attack Code Raises Stakes for USB Security*. Available at: https://threatpost.com/badusb-attack-code-publicly-disclosed/108663/ (accessed on 2 October 2014).

9

Composable Framework Support for Software-FMEA through Model Execution

Valentina Bonfiglio[1], Francesco Brancati[1], Francesco Rossi[1], Andrea Bondavalli[2,3], Leonardo Montecchi[2,3], András Pataricza[4], Imre Kocsis[4] and Vince Molnár[4]

[1]Resiltech s.r.l., Pontedera (PI), Italy
[2]Department of Mathematics and Informatics, University of Florence, Florence, Italy
[3]CINI-Consorzio Interuniversitario Nazionale per l'Informatica-University of Florence, Florence, Italy
[4]Dept. of Measurement and Information Systems, Budapest University of Technology and Economics, Budapest, Hungary

9.1 Introduction

Performing Failure Mode and Effects Analysis (FMEA) during software architecture design is becoming a basic requirement in an increasing number of domains. However, due to the lack of standardized early design-phase model execution, classic Software-FMEA (SW-FMEA) approaches carry significant risks and are human effort-intensive even in processes that use Model-Driven Engineering (MDE).

From a dependability-critical development process point of view, FMEA – more generally, the identification of hazards and planning their mitigation – should be performed in the early phases of system design; for software, this usually translates to the architecture design phase [1]. Additionally, for some domains, standards prescribe the safety analysis of the software architecture – as is the case, e.g., with ISO 26262 in the automotive domain.

However, historically, software architecture specifications in the most widely used modelling languages either do not represent behaviour, only structure, or the behavioural models do not have standardized operational semantics. This is a major problem for SW-FMEA; in contrast to hardware, relatively small changes of "internals" of a software component (essentially the program logic) can lead to wide variations in the response of executed software components to various external and internal faults. This means that in addition to computing error propagation from component to component, the sensitivity of each component for internal and external faults has to be explored on a case by case basis, and this can be done only by using specifications of behaviour.

In the absence of this capability, the system modeller has to either make strong guarantees in advance ("this component will be fail-silent under all circumstances"), or make too pessimistic assumptions (e.g., "all kinds of output failures are possible"). Significant risk is introduced by the fact that the error propagation assumptions made at this stage have to hold for the final system – otherwise the constructed hazard mitigation arguments will not hold, either. Thus, without rolling back the development process, we run the risk of having to enforce not easily enforceable guarantees, or having to use dependability mechanisms that are actually superfluous in the given system.

This chapter addresses the aforementioned problem on the basis of a new standard for the UML 2 modelling language. Throughout the next sections, we will introduce the reader to advances in standardized model execution semantics, the outline of a composable framework built on top of executable software architecture models to help SW-FMEA, as well as a realization of such a framework applied on a case study from the railway domain.

9.2 Software-FMEA Using fUML/ALF

For UML 2, the status quo of not having standardized operational semantics has changed with the standardization of "Foundational UML" (fUML) [2]: a core subset of UML 2 has been given standardized execution semantics. Although fUML mainly contains facilities for describing structured activities of communicating, typed objects, in theory, the whole UML 2 language can be mapped to it. To facilitate practical application, fUML also has an action language called Alf, the "Action Language for Foundational UML" [3].

Alf is a quasi-imperative, Java-like programming language. As a "surface language" for fUML, its structure and execution is directly and unambiguously mapped to fUML. Whole programs can be written purely in Alf, but it can be also used to define specific behaviours in an encompassing UML 2 model. However, in this case, the operational semantics of the embedding model containing the Alf code snippets also has to be specified, e.g., by translating the whole model to pure Alf. This is not necessarily a shortcoming; our approach actively exploits the partially “underdefined” composite structure semantics. That said, it is worth to note that the newly finalized standard “Precise Semantics of UML Composite Structures” (PSCS) [4] addresses exactly this issue.

9.2.1 Tooling for fUML and Alf

Due to the novelty of the languages, fUML and Alf tooling is still maturing, but the progress is steady. For both fUML and Alf, reference interpreters exist [5]; for fUML, additional execution engines are also available [6]. Papyrus, the popular Eclipse-based modelling environment includes an Alf editor for UML 2 language elements and supports direct compilation of Alf code into UML 2 [7].

The compilation of fUML/Alf to other languages and the formal analysis of fUML/Alf specifications are much less developed, with no directly (re)usable solutions known. That said, notably [8] presents a full Model-Driven Engineering approach where Alf code is translated into an intermediate model that, in turn, is translated to C++. On the formal analysis side, initial progress has been made both for theorem proving [9] and classic model checking [10].

9.2.2 Software-FMEA through Alf Execution

Earlier work performed in the CECRIS project (“Certification of Critical Systems”) [11] has proposed an approach for the SW-FMEA of component-based systems through Alf execution (using an interpreter) [12]. The main idea of the approach is summarized in the following three steps.

1. Components as well as their Alf code are translated into a single Alf program. During translation, the code for a cyclic scheduler component is also woven into the Alf source (with a simple logical clock). The static component activation schedule is determined by the modeller.
2. As a form of model-level fault injection, the translation can inject some simple errors into the scheduler as well as replace output/input port behaviours with “programmed” error behaviours.
3. Error propagation is analyzed by comparing simulation runs of the fault-free case to various fault activations.

In general, simulation certainly has its drawbacks; e.g., it is hard to ensure that all execution paths have been exercised in a nondeterministic system, though this is not a major issue for three reasons. First, the reference simulator performs sequential execution with deterministic choices (a semantic variation that the fUML standard fully allows). Second, although the embedded system models we apply our approach to do not exhibit parallelism at either the micro or the macro level, there is at least one fUML virtual machine called *moliz* that supports very fine-grained external control of model execution [13]. This means that if the need arises, the various interleaving executions can be tracked, accounted for and controlled. Thirdly, we do expect solutions for the application of formal methods (at least model checking) on fUML/Alf models to appear in the near future; these, by their nature, cover the entire state space of models.

These considerations demonstrate an important strength of the approach proposed in Bonfiglio et al. [12] and provides one of the main motivations for the framework presented in this chapter. If, during the translation of the component model to pure Alf, we are able to equip the Alf code with all the facilities that transform model execution into explicit error propagation execution, then we can reap the benefits of advances in fUML/Alf tooling without additional effort.

9.2.3 Framework Support for Executable Error Propagation

Along the previous consideration, we describe the design of a model transformation framework that transforms component models with Alf behaviour specifications into a pure Alf program that simulates error propagation by passing error tokens between the components and computing (or approximating) the input-output error transformation that a potentially faulty software component exhibits upon (erroneous or correct) activation (Figure 9.1). As the user-supplied Alf code cannot always be used to compute error propagation (e.g., the component itself might have an active internal fault), in some cases, error transformation draws on a library of behavioural patterns (e.g., "fail-silent").

The orchestration of the execution of components is broken into a number of configurable, cooperating functions. These functions have generic variants; these are drawn from a framework library of options (Figure 9.2).

9.2.4 Error Tokens, Component Activation

The composite error tokens passed between components carry a reference value – the object that should be seen during the interaction in a fault free system – as well as error information. The error being passed is either a standard category (succinctly introduced, e.g., in TanjaMayerhofer [14]), a refinement thereof, or a specific one (e.g., a specific erroneous value that is late by a known amount of time).

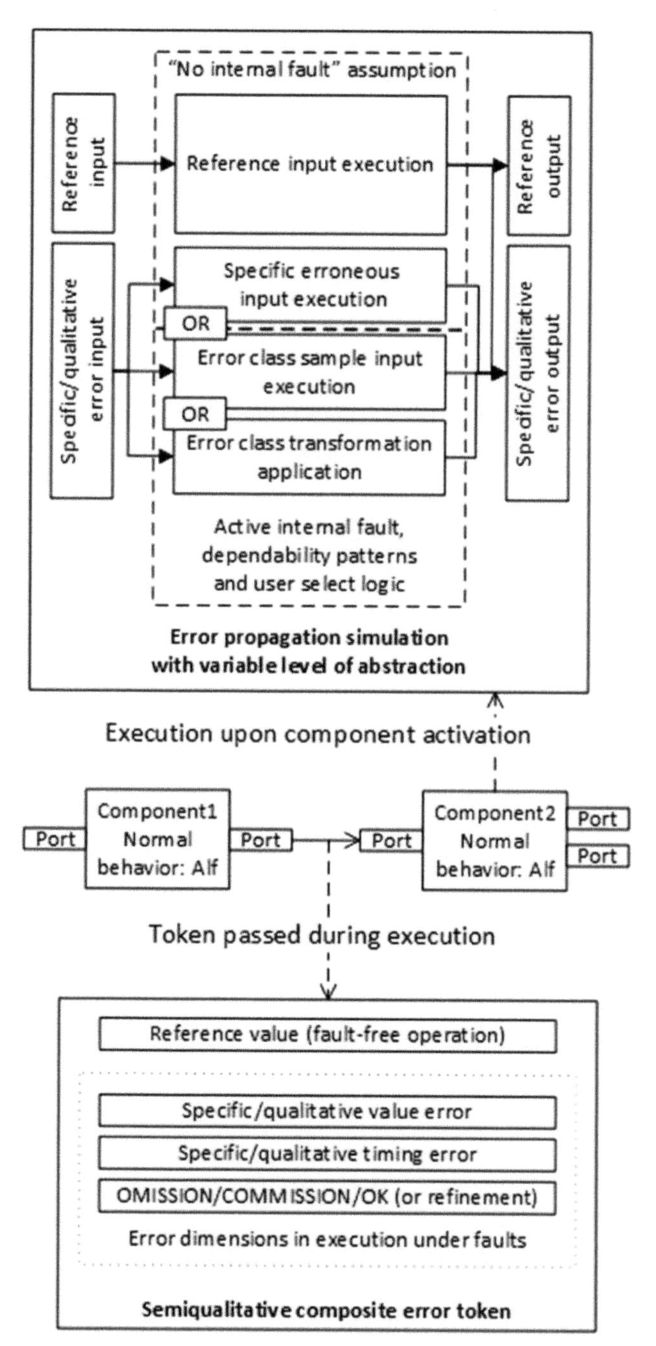

Figure 9.1 Composite error token passing during execution and component activation.

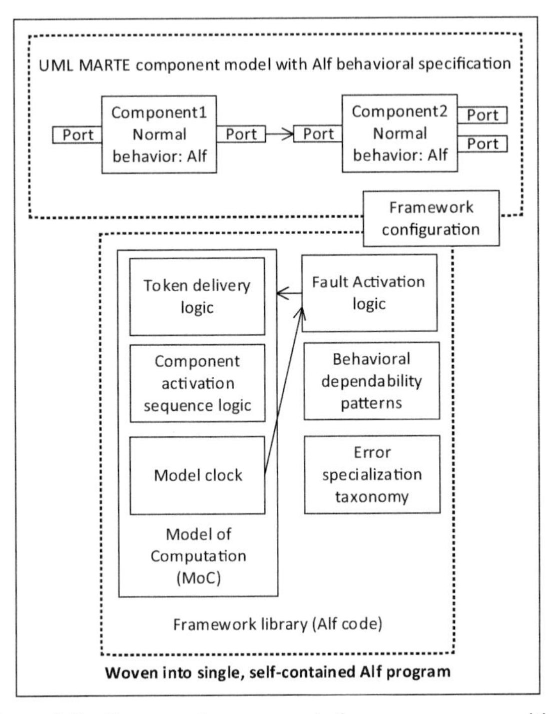

Figure 9.2 Framework components for program composition.

Component activation computes the output error tokens of the component based on the input ones. The logic for producing the error outputs depends on numerous, but mostly straightforward factors (Figure 9.2); to note is that for computing the error output when the specific error is not known, only the category, the modeller may decide to either run the user-supplied Alf code on a sample from the class or use a predefined error category transformation logic from a library of dependability behavioural patterns.

9.2.5 Execution Orchestration

Component models in UML 2 do not have standard execution semantics; the cyclic execution logic with a static schedule in Bonfiglio et al. [12] (summarized in Section 9.2.2) came from the domain of the modelled system. As a matter of fact, the overall approach is able to support numerous models of computation (MoC) – rules defining the semantics of control, concurrency, communications and time. Synchronous data flow networks, discrete events, static scheduling, and workflow-like execution all fit the approach through configurable, reusable implementations in Alf (with varying complexity, of course). In order to be able to account for orchestration errors, the framework is also intended to support runtime fault injection on the MoC implementations.

9.2.6 Fault Injection

Fault injection is performed by configurable fault activation logic implementations. These determine active faults of the various components (including orchestration) at various points in time (if the MoC defines a notion of time).

9.3 Case Study: Application of Software-FMEA through Model Execution

The case study used for the Software FMEA process was based on the railway domain, more specifically the European Rail Traffic Management System (ERTMS) and its Control Command part European Train Control System (ETCS) [15]. ERTMS/ERTCS is an automatic train protection system, and as such, a safety-critical system. ERTMS is composed of trackside units (e.g. beacons for positioning and information reporting) and on-board units. The full system is rather complex, thus in the case study only a small, simplified part of the specification was modelled. The focus of the modelling was on the safety function of receiving and consistency checking of messages from trackside beacons called *balises*.

9.3.1 Definition of the Modelled System

The case study system was based on the balise-related basic functionality of ERTMS/ETCS. The system was modelled using UML and Alf (Action Language for Foundational UML) [16]. The static structure part of the system was also described with the textual syntax of Alf; however, some of these will be represented on graphical diagrams for convenience.

As the envisioned Software-FMEA approach should be applied early in the design phase, no actual implementation or detailed design model is available in this stage. Therefore, in order to exercise the behaviour defined in the executable model, the "environment" of the modelled system had to be simulated. This scaffolding was also developed in Alf, thus the model consists of two main parts.

- **Target system**: the On-board Unit (OBU) of ETCS, the core software running in the train.
- **Environment**: simulation of the track, trackside equipment and movement of the train.

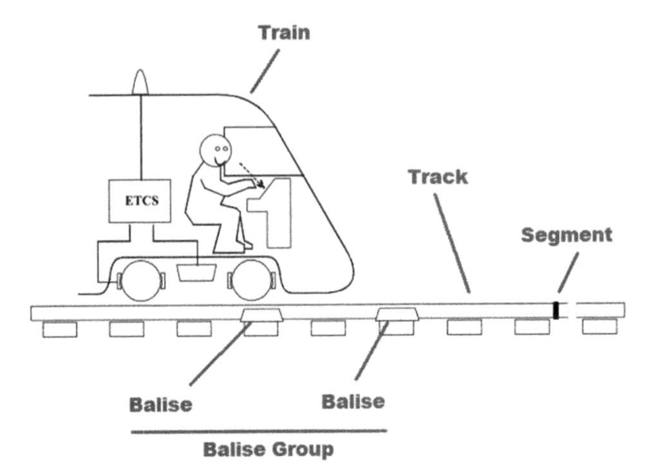

Figure 9.3 Parts of the simulated environment in the case study (figure based on European Railway Agency [15]).

Note that to reduce the complexity, the simulation of the environment focuses only on the necessary details to support the modelled functionality. Hence, the simulation is based on discrete events, and speed, distance and braking are all abstracted.

The main elements in the environment of the system are depicted on Figure 9.3 (based on Figure 2.6 in ETCS Subsection 026 Chapter 2 [15]) and are briefly explained below.

- *Track.* The train is moving on a track (the actual physical dimensions of the track are abstracted in the case study).
- *Segment.* The track is composed of neighboring segments. The train can move from one segment to another neighboring one.
- *Train.* The train can move in forward or backward one segment in each simulation step (the actual speed of the train is abstracted).
- *Balise.* A passive beacon deployed onto the track. When the train passes over a balise, it powers it up remotely via radio waves, causing the balise to send a so called telegram to the train.
- *Balise Group.* Balises can be organized in balise groups. A balise group can contain up to 8 balises. By giving position numbers to individual balises inside a group, the train can identify direction and detect missed balises.

The modelled target system consists of the main parts depicted on Figure 9.4 and explained below.

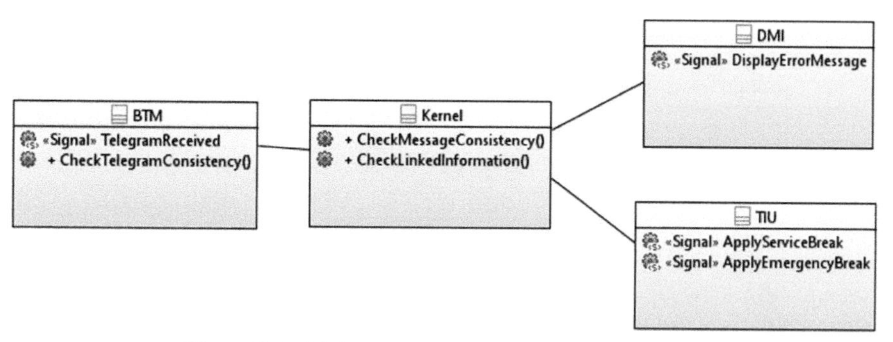

Figure 9.4 Main components of the modelled system.

- *Balise Transmission Module* (BTM). Responsible for receiving raw, individual telegrams from the balises, checking and then forwarding them to the Kernel.
- *Kernel*. Responsible for the core functionality in ETCS. In the current case study, it collects telegrams from different balises to form and analyze balise group messages. If an error is detected, it can notify the driver through DMI or control the train through TIU.
- *Driver Machine Interface* (DMI). Can display information on the driver interface.
- *Train Interface Unit* (TIU). Can control the train. In the current case study, it can apply breaking.

The simulated environment and the target OBU system is connected by sending and receiving balise telegrams. The structure of a telegram is defined in Chapter 8 of the ETCS Subset 26 (SRS) [15], and is summarized in Figure 9.5.

The telegram itself was modelled using data types in Alf. The simulated balise and the BTM module directly work on this data structure, while the Kernel receives an object structure built by the BTM based on the telegrams.

The modelled behaviour is attached to the active classes in the system. Basically, they are all waiting for signals to receive, and then perform the signal handler behaviour specified in Alf. For example, upon receiving a raw telegram, the BTM checks the consistency of the header fields. This was implemented in the Alf activity presented in Figure 9.6.

The model and the modelled scenarios were executed in the Alf Reference Implementation [17]. The model includes logging to create execution traces. For example, the output in Figure 9.7 shows a simple, valid execution trace where the OBU receives a consistent telegram from a single balise. The same

GENERAL FORMAT OF BALISE TELEGRAM			
Field No.	VARIABLE	Length (bits)	Remarks
1	Q_UPDOWN	1	Defines the direction of the information: Down-link telegram (train to track) (0) Up-link telegram (track to train) (1)
2	M_VERSION	7	Version of the ERTMS/ETCS system.
3	Q_MEDIA	1	Defines the type of media: Balise (0)
4	N_PIG	3	Position in the group. Defines the position of the balise in the balise group.
5	N_TOTAL	3	Total number of balises in the balise group.
6	M_DUP	2	Used to indicate whether the information of the balise is a duplicate of the balise before or after this one.
7	M_MCOUNT	8	Message counter (M_MCOUNT) – 8 bits. To enable detection of a change of balise group message duringpassage of the balise group.
8	NID_C	10	Country or region.
9	NID_BG	14	Identity of the balise group.
10	Q_LINK	1	Marks the balise group as linked (Q_LINK = 1) or unlinked (Q_LINK = 0).
	Packet 0 (optional)	14	Virtual Balise Cover marker.
	Information	Variable	This information is composed according to the rules applicable to packets.
	Packet 255	8	Finishing flag of the telegram.

Figure 9.5 Structure of a balise telegram [15].

```
privateactivityCheckTelegramConsistency(in t : Telegram) : Boolean {
let consistent: Boolean = true;
if (t.Q_UPDOWN != UpDown.Up || t.Q_MEDIA != Media.Balise ||
t.N_PIG<0 || t.N_PIG>7 ||
t.N_TOTAL<0 || t.N_TOTAL>7) {
        consistent = false;
    }
//further checks ...
return consistent;
}
```

Figure 9.6 Alf implementation of a BTM behaviour.

```
  [test   ] SingleBalise_Valid_ReceiveTelegram
[train ] Received MoveForward
[train ] Moved to segment s2
[train ] train   -> s2      : TelePower
[s2     ] Received TelePower from train
[s2     ] s2      -> b1      : TelePower
[b1     ] Received TelePower from train
[b1     ] b1      -> BTM     : TelegramReceived
[BTM    ] Received Telegram from Balise with position 0 in BG 2
[BTM    ] BTM     -> KERNEL : TelegramReceived
  [KERNEL] Received Telegram from Balise 1 in BG 2, consistent:
true
```

Figure 9.7 Log trace of a fault-free execution of the case study model.

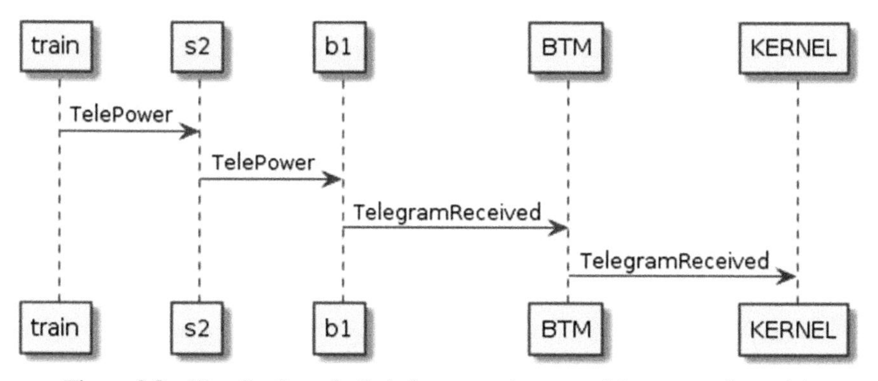

Figure 9.8 Visualization of a fault-free execution tree of the case study model.

trace can be visualised by PlantUML as a sequence diagram (Figure 9.8). In the modelled environment, there is a train, a track with two segments (s1, s2) and a balise (b1). The train initially stands on the first segment (s1) and it moves to the second (s2) as the first step of the test case. Note that for the sake of simplicity, the component representing the train in the case study is not associated with the balise component, so powering up a balise is mediated by the segment component.

9.3.2 Process Evaluation

As the discussion in section "Software-FMEA Using fUML/ALF" pointed out, the fundamental tenet of our method is to perform SW-FMEA on component-based systems through Alf execution (as of now, using an interpreter).

Based on the presented use case, process-wise, it is apparent that the SW-FMEA approach can be used in a "drop-in" fashion in existing safety processes, replacing classic approaches during software architecture analysis. The major added value is delivering much tighter bounds on error propagation characteristics (certainly not probabilities!) at the point in design where the major dependability mechanisms are most probably decided upon. While much more sophisticated than classic FMEA (and even such composable methods as HiP-HOPS [18]), the approach largely remains an FME(C)A – and thus there is no real reason it cannot be a candidate method in virtually all safety processes where SW-FMEA is necessary.

Importantly, the ability to "mix and match" specific errors and error categories in evaluating and propagating errors may enable new process patterns. Refinement of our knowledge of the error propagation characteristics in the system is a definite (and largely new) option in this setting; thus, in theory, safety arguments could very well evolve *cooperatively* with the refinement of system and software design. Future research will explore this possibility.

Certainly, there are some apparent drawbacks, too.

- **Modelling overhead**. The least significant drawback that nevertheless has to be mentioned is that the whole approach assumes that the system under design is created in an appropriate Model-Driven System Design (MDSD) workflow. Although MDSD is becoming the default in many industries where SW-FMEA has to be performed, it is not necessarily used in all settings.
- **Early definition of behaviour**. Executable models such as Alf give us the possibility to model behaviour early on in the design process – but this does not automatically mean that it is convenient or feasible at all. Further studies are necessary to evaluate this aspect.
- **Proof of behavioural equivalence**. When executable behavior is specified early on in the development process and it is the basis of safety arguments, behavioural equivalence of the final system (and components) with this early specification has to be maintained during development.
- **Simulation.** As of now, we use simulation for model evaluation. Simulation has its drawbacks; e.g., it is hard to assure that all execution paths have been exercised in a nondeterministic system. We argued in chapter "Software-FMEA through Alf execution" that in our case this is not a major issue. In fact, the proposed approach does not rely on any specific simulation technique; all the facilities that transform model execution into explicit error propagation execution are included in the model. This way, we will be able to reap the benefits of advances in fUML/Alf tooling without additional effort.

9.4 Implementation in a Blockly-based Modelling Tool

To demonstrate the general applicability of the approach presented so far, the main points of the framework were also implemented for the modelling tool introduced in "Chapter 4 – SYSML-UML like modeling environment based on Google Blockly customization".

The tool supports the modelling of static and dynamic aspects of component-based systems by using blocks, interfaces and connections to model structure, as well as sequence diagrams to model the collective behaviour of the whole system. The former aspect defines the participants and their relations, while the latter describes their interactions and the exchanged data. The basic block of behaviour is a *Service*, which may have a specific implementation in Python. Interactions then consist of *Service Requests* and control logic (e.g. decisions). Once modelled, the tool can visualize the connections in the system, as well as the sequence diagram defined for the global behaviour. One of the strongest aspects of the tool is the ability to generate a Python program for the simulation of the system. With small modifications, the generated code is an appropriate candidate for the methods defined in the previous sections.

9.4.1 Preparation of the Model

The case study model was again based on the balise-related basic functionality of ERTMS/ETCS (Figures 9.3 and 9.4 for the structure and Figure 9.7 for the behaviour). A bird's eye view of the model itself is presented in Figure 9.9.

The generated code has been augmented with logging: values of parameters and variables after assignment result of decisions and assertions as well as service requests were all output to build an execution trace.

As before, faults activation was done by injected, configurable logic that would determine which faults are activated during execution. In practice, this involves the replacement of certain constructs (e.g., expressions) with a function call that either performs the original behaviour (e.g., returns the value of the original expression) or alters the behaviour in some way (e.g., negates a condition). In the current case study, faults affected the assignment of Boolean and integer values (altering the value of the right-hand-side expression), the conditions in decisions and the sending of service requests (causing an *omission* fault).

The fault activation logic can be fine-tuned by setting the maximum number of active faults as well as if the faults are transient of permanent. In case of permanent faults, fault distribution is balanced by an initialization logic that randomly selects a configured number of faults, which may then activate if

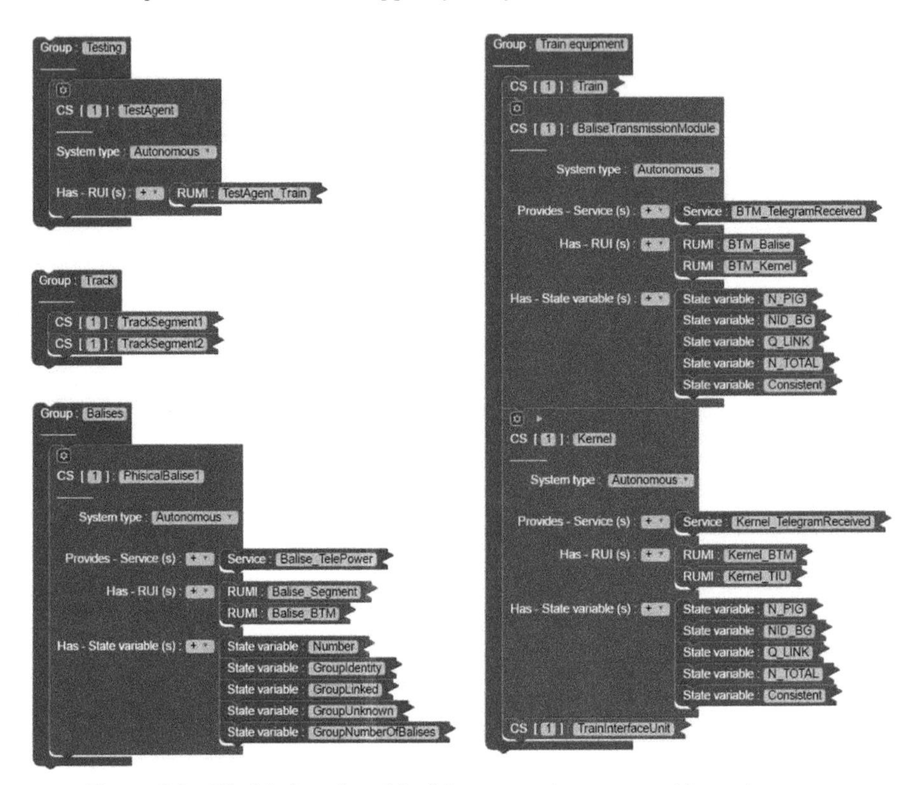

Figure 9.9 Blockly-based model of the case study system and its environment.

the affected statements are executed (i.e. the injected fault activation function is called). In this case study, the logic was configured to activate *at most one* permanent fault.

Faults were injected in the relevant parts of the control logic of the train (i.e. the *Kernel* and the *Balise Transmission Module*), but message omissions were also included in the code simulating the powering of the balise to emulate failure of equipment. Every time a fault activated, its type and the affected line were logged, but not the specific value used to modify the original expression.

Two test cases were used for the simulations: in the first one, the balise sends a consistent telegram, while in the second; the balise has corrupted data (it has an invalid position value). Thus, according to the specified behaviour, correct reactions of the system would be to acknowledge the reception of the telegram in the first case, and applying emergency brake in the second. Assertions in the model checked if the produced behaviour was in accordance with the balise data, as well as if a telegram was successfully processed in a given time frame (in more complex settings, this could be detected and handled when reaching the next balise).

9.4.2 Aggregation and Analysis of Traces

A single simulation of the fault-free model and 1000 simulations with random faults in each test case provided a satisfying number of traces to conduct a probabilistic analysis. The traces were processed through the following steps:

1. *Error traces:* Every faulty trace was compared to the reference (fault-free) trace to obtain the differences, i.e. to identify the chain of errors that led from a single fault activation to a failed assertion. Corresponding to the injected faults, the errors could be *Parameter errors*, *Data errors*, *Control errors* and *Missing calls.*
2. *Superposition of traces:* The error traces were merged to obtain a graph. An arc in this graph from A to B means that in some trace, error A was immediately followed by error B.
3. *Annotation with occurrences:* Arcs of the graph were then annotated by the ratio of the number of traces in which B has *eventually* occurred after A to the total number of traces in which A has occurred. This value corresponds to the *conditional probability* of eventually seeing B if A has occurred. This way, a probability of 1 means there is a strong correlation between errors A and B, which in this case may also suggest a causal relationship. Hence, these cases are visually distinguished by using a solid line for the potential causalities and a dashed line otherwise.
4. *Reducing noise:* Various techniques were employed to remove arcs that were the consequences of other relationships. It is worth to note that, this part of the process is the most theoretic and has a lot of room for improvements. The more efficient the techniques used here are, the more meaningful the results of the process can be.

Nodes of the graph (fault activations, errors and assertion failures) were grouped by the component which logged them. The output for the test case with the valid telegram can be seen on Figure 9.10. Things to node about the figure are the following.

- In an FMEA terminology, each box corresponding to a component contains the internal faults and the failure modes (errors) of the component. Arcs entering the box denote external failure modes that affect this component, while outgoing arcs denote failure modes of the component that affect others.
- In the case study model, omission of service requests always results in a failure to handle the balise.

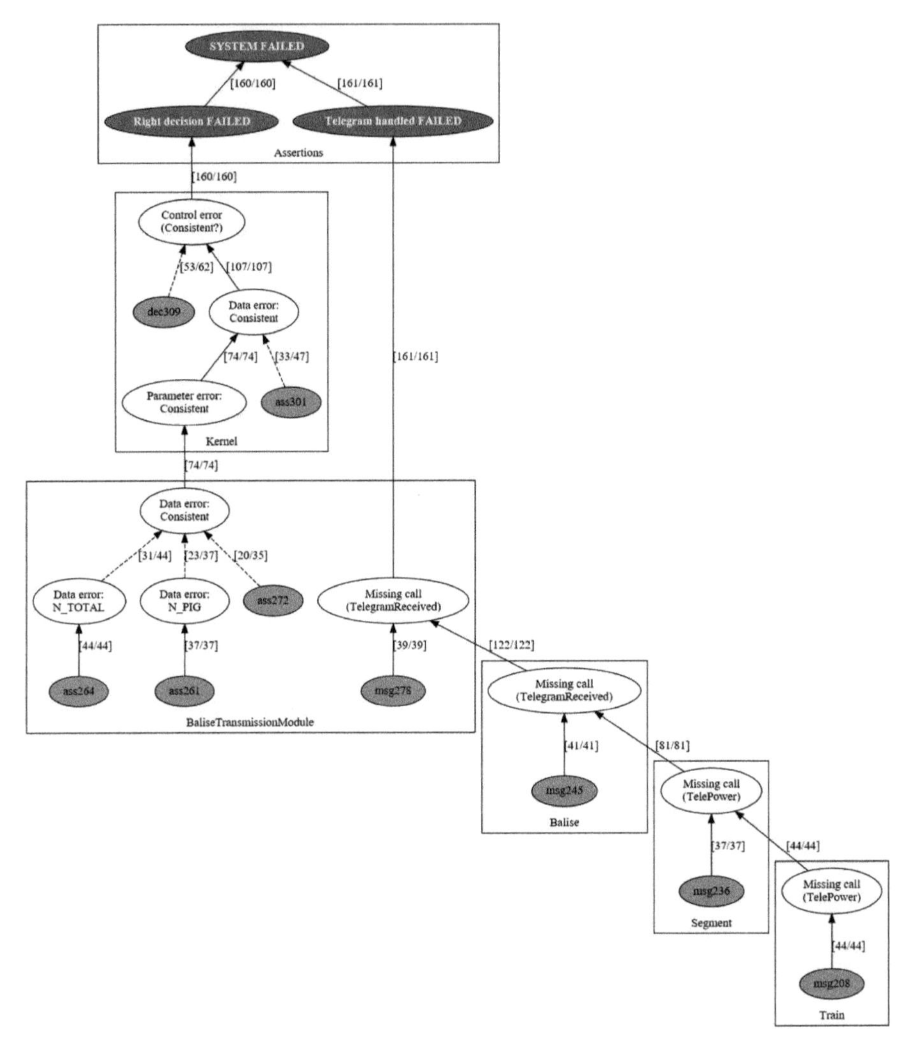

Figure 9.10 Error propagation in the case study model when input is consistent.

- On the other hand, corruption of data causes a system-level failure only if the value of the Boolean flag "Consistent" gets corrupted. Once this happens, there is no way to avoid failure, but only some corruption of the other values leads to the corruption of the flag.
- A fault affecting the value of "Consistent" does not always cause a data error, because always returning *True* is considered a correct answer in this scenario.

Analysis of the test case with the invalid telegram showed similar results.

9.5 Concluding Remarks

In the chapter, the reader was introduced to the main ideas of a novel approach to SW-FMEA for component-based systems that can be composed with existing safety processes. The method can replace or augment classic approaches during software architecture analysis and automating much of the traditional FMEA techniques.

The work transfers techniques well-known in academia into the SW-FMEA of safety-critical embedded systems, with strong potential applicability in other dependability-critical domains. These techniques include explicitly embedding fault activation logic and operational semantics into the interpreted model and constructing error automata from the specification of normal and abnormal behaviours (see e.g., [19]). At the same time, the presented approach promises to have a low effort overhead over producing the base models (that are produced in a model-driven process even in the absence of SW-FMEA); something that is sorely missing from manually performing SW-FMEA.

References

[1] Pataricza, A. (2007). "Systematic Generation of Dependability Cases from Functional Models," in *Proceedings of the Symposium FORMS/ FORMAT – Formal Methods for Automation and Safety in Railway and Automotive Systems*, Budapest, Hungary.

[2] Object Management Group. (2016). *Semantics of a Foundational Subset for Executable UML Models (fUML), version 1.2.1.*

[3] Object Management Group. (2013). *Action Language for Foundational UML (Alf), version 1.0.1.*

[4] Object Management Group. (2015). *Precise Semantics of UML Composite Structures (PSCS), version 1.0.*

[5] GitHub. (2016). *Foundational UML Reference Implementation.* [Online]. Available at: https://github.com/ModelDriven/fUML-Reference-Implementation (accessed on 1 February 2016).

[6] GitHub. (2016). *moliz – Model Execution Based on fUML* [Online]. Available at: https://github.com/moliz (accessed on 1 February 2016).

[7] Seidewitz, E, and Tatibouet, J. (2015). "Combining Alf and UML in Modeling Tools – An Example with Papyrus," in *OCL 2015 – 15th International Workshop on OCL and Textual Modeling: Tools and Textual Model Transformations Workshop Proceedings*, 105–119.

[8] Ciccozzi, F. (2014). *From Models to Code and Back: A Round-trip Approach for Model-driven Engineering of Embedded Systems*. Doctoral thesis, Mälardalen University, Sweden.

[9] Romero, G., Schneider, K., and Ferreira, M. G. V. (2014). "Using the base semantics given by fUML for verification," in *2014 2nd International Conference on Model-Driven Engineering and Software Development (MODELSWARD)* (New York, NY: IEEE), 5–16.

[10] Schneider, A. S. and Treharne, H. (2011). "Towards a Practical Approach to Check UML/fUML Models Consistency Using CSP," in *Formal Methods and Software Engineering*, eds S. Qin and Z. Qiu (Berlin: Springer), 33–48.

[11] CECRIS. (2016). *CECRIS – Certification of Critical Systems, Grant Agreement no.: 324334, IAPP Marie Curie Action, 7th Framework Program*. Available at: http://www.cecris-project.eu (accessed on 16 January 2016).

[12] Bonfiglio, V., Montecchi, L., Rossi, F., Lollini, P., Pataricza, A., and Bondavalli, A. (2015). "Executable Models to Support Automated Software FMEA," in *2015 IEEE 16th International Symposium on High Assurance Systems Engineering (HASE)* (New York, NY: IEEE), pp. 189–196.

[13] Object Management Group. (2011). *UML Profile for MARTE: Modeling and Analysis of Real-Time Embedded Systems, version 1.1*.

[14] Mayerhofer, T. (2014). *Defining Executable Modeling Languages with fUML*. Doctoral thesis, Vienna University of Technology.

[15] European Railway Agency. (2014). "ERTMS/ETCS System Requirements Specification", SUBSET-26.

[16] Object Management Group. (2013). Semantics of a Foundational Subset for Executable UML Models (fUML).

[17] ModelDriven. (2016). ModelDriven.org: Action language for UML (Alf) open source implementation. Available at: http://modeldriven.github.io/Alf-Reference-Implementation/

[18] Papadopoulos, Y., McDermid, J., Sasse, R., and Heiner G. (2001). Analysis and synthesis of the behaviour of complex programmable electronic systems in conditions of failure. *Reliabil. Eng. Syst. Safety* 71, 229–247.

[19] Gallina, B., and Punnekkat, S. (2011). "FI4FA: A Formalism for Incompletion, Inconsistency, Interference and Impermanence Failures' Analysis," in *2011 37th EUROMICRO Conference on Software Engineering and Advanced Applications (SEAA)* (New York, NY: IEEE), 493–500.

10

A Monitoring and Testing Framework for Critical Off-the-Shelf Applications and Services

Nuno Antunes[1], Francesco Brancati[2], Andrea Ceccarelli[3,4], Andrea Bondavalli[3,4] and Marco Vieira[1]

[1]CISUC, Department of Informatics Engineering, University of Coimbra, Portugal
[2]Resiltech s.r.l., Pontedera (PI), Italy
[3]Department of Mathematics and Informatics, University of Florence, Florence, Italy
[4]CINI-Consorzio Interuniversitario Nazionale per l'Informatica-University of Florence

One of the biggest verification and validation challenges is the definition of approaches and tools to support systems assessment while minimizing costs and delivery time. Such tools reduce the time and cost of assessing Off-The-Shelf (OTS) software components that must undergo proper certification or approval processes to be used in critical scenarios. In the case of testing, due to the particularities of components, developers often build *ad hoc* and poorly-reusable testing tools, which results in increased time and costs. This chapter introduces a framework for testing and monitoring of critical OTS applications and services. The framework includes (i) a box instrumented for monitoring OS and application level variables, (ii) a toolset for testing the target components, and (iii) tools for data storing, retrieval and analysis. We present an implementation of the framework that allows applying, in a cost-effective fashion, functional testing, robustness testing and penetration testing to web services. Finally, the framework usability and utility is demonstrated based on two different case studies that also show its flexibility.

10.1 Introduction

Verification and Validation (V&V) has been largely applied in scenarios that involve life and mission critical embedded systems, and is dominantly used as a design-time quality control process for the purpose of evaluation of the compliance between of a product, service, or system [1]. Checking a system using traditional V&V methods frequently exceeds the effort needed for the core development time. In fact, rigorous V&V in on the fundaments of critical applications and has been applied in several domains as the railway [2] and space [3], and recently a strong effort has been made to standardize these practices for automotive [4].

Although the industry rapidly turns to system integration based on the reuse of hardware and software components, also known as off-the-shelf (OTS) components, it is still necessary to apply rigorous V&V techniques to assess the applications. While hardware OTS are nowadays widely accepted, and used (they have their own certification), software OTS still creates serious difficulties to companies, which are on one hand constrained to meet predefined quality goals, whereas, on the other hand, are required to deliver systems at acceptable cost and time to market. Large companies mainly follow a brute-force approach by focusing large volume investment into tooling and in-house training, but even high-tech SMEs are highly vulnerable to the new challenges.

In this context, one of the biggest challenges to the V&V community is to define **methods, strategies and tools able to validate a system adequately**, while simultaneously keeping the **cost and delivery time reasonably low**. The key part of the challenge is to establish a proper balance between achievable quality with a particular technique (in terms of RAMS attributes) and the costs required for achieving such quality. The problem grows when it is necessary to include COTS components in a critical system that must be certified. As a matter of fact, although modern standards consider the possibility of assessing products, which encompass COTS software, this is still considered a challenge [5].

In industrial practices, *integration* and *usage* of OTS software components in critical systems is generally supported by two different assessment processes, both to understand the behaviour of the component and to assess that it does not introduce hazards in the system. In the first, whenever applicable, the activity is limited to assess the integration, verifying that the OTS component is properly wrapped in the system without affecting system's safety. In the second, a complete assessment of the OTS component is performed; this may include activities as production of documentation, reverse engineering, and static analysis, among others. For companies, this usually means a reasonable amount of **effort in developing a specific tool** that can support the testing of a specific OTS component to be integrated in a certain critical system.

This chapter presents a framework for testing and monitoring critical applications and services. The framework monitors the variables of the system while applying diverse forms of testing over the applications. This way, it is possible to better detect problems in the applications as well as better diagnosing them, maximizing the effectiveness of the tests. The framework is based on an application independent and reusable core infrastructure, allowing the user to apply cost effective practices. The proposed framework consists of two main components, as follows.

The first, named **Instrumented System** is a monitoring environment where the applications or services can be executed and monitored. The kernel of the operating system is instrumented to monitor all the variables that are representative for V&V process. The environment also includes middleware that is also instrumented to provide values of the all the variables representative for V&V at this level. The second component named **Test and Collect** contains a set of tools for application testing and, data storage and analysis. The testing tools included should be able to generate different types of testing including functional testing, robustness testing, security testing, etc. For data storage the framework includes a database management system and tools to allow the user, in a semi-automated way, to generate a schema able to store the values of the monitored variables.

The use of this kind of analysis is essential for the conscious use of OTS components. By testing the OTS, it is also possible to use wrapping strategies [6, 7] around the identified problematic parts of the component. An important part of the implementation is that one instance of this component can be connected to multiple instrumented systems. This way, the framework is prepared to be extended for other purposes, as in the case of monitoring a large scale system with multiple nodes, as it is possible to correlate data from multiple sources and also analyse more complex systems.

Several works have shown the usefulness of system monitoring to detect anomalies in the system. Statistical analysis algorithms have been used in the past for on-line fault detection [8]. This technique overcomes some of the limitations of static threshold analysis, that for instance in [9] monitoring techniques are used to detect application hangs. Works towards certifying OTS components are also not new. The technique [10] tries to determine the quality of OTS components using black box and fault injection in two phases: first, the component is tested to make sure it works properly, and second, the system is tested to make sure that the system works even if the component presents an incorrect behaviour.

The chapter also presents a prototype implementation and demonstration of the framework. The implementation includes tools that allow the user to apply to the web services different types of testing: functional, stress, robustness and penetration testing. During the different testing processes, the system variables are monitored both at middleware level and at operating system level. Two different case studies were devised to demonstrate and evaluate the framework.

The first **case study is focused on the services of the Life ray Portal**, an enterprise web platform project that aims for immediate delivery of robust business solutions for organizations. This case study allowed us to demonstrate the flexibility, usability and utility of the framework. The results revealed the services under test performing quite well in the situations tested. Obviously, the quality of the tests performed depends on the testing tools used, but this discussion is out of the scope of this work, as the merits of each tool were evaluated and discussed in different works by their authors [11, 12].

The second **case study is focused on simulator of a railway** environment that includes a system that should detect anomalous and hazardous situations on the trains running on that line. The stringent requirements of the system that should be tested and validated exactly in the same setup as it will operate demonstrated the flexibility of the toolbox, which was able to be easily ported into such environment.

The chapter is organized as follows. Section 10.2 describes the concepts behind the framework, while Section 10.3 presents the implementation details. Section 10.4 presents the case studies used to demonstrate the framework and the respective details. Section 10.5 concludes the section and puts forward relevant future work.

10.2 Framework Architecture

Our proposal is an advanced framework for testing and monitoring critical applications and services. Despite the most common approach for testing OTS web services is the "black box", the tool has been designed to take advantages of any piece of information available. The overall architecture of the proposed framework is depicted in Figure 10.1. As it is possible to observe, the framework is based on two main components: i) *Instrumented System*, which the system in which the web service is running, and ii) *Test and Collect*, which is used to stimulated the web service and to collect evidences

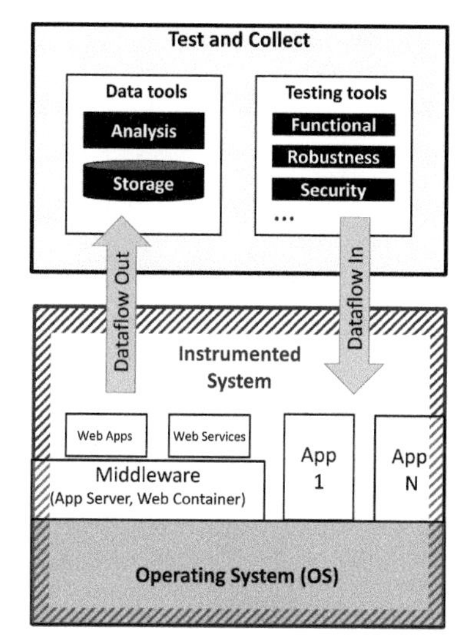

Figure 10.1 Framework architecture: overall view and interactions.

of its behaviour. Although the current implementation focuses on Java Web Services running over a Tomcat 7 Application Server (AS) and a Linux CentOS 6 Operative System (OS), the proposed solution can be evolved to different platforms and Web Services Middleware (AS).

As we can observe, Figure 10.1 also shows the interactions between the two systems of the framework: the testing tool invokes methods of the web service triggering specific functionalities, and at the same time the analysis tools read information on the overall status of the operative system and service middleware. The next sections present the concepts behind each component.

10.2.1 Instrumented System (IS)

The Instrumented System is a monitoring environment where the applications or services can be executed and tested. Considering that weaknesses can affect the middleware layer (e.g., depleting available free memory in the heap) and the operating system layer (e.g., exchanging a huge amount of data or delaying the overall system), both are object of monitoring.

Figure 10.2 represents the monitored components (Operating System, Middleware, Applications) and data flows. The key innovation is to monitor both the variables of the *OS* and of the *Middleware* (when applicable) at the same time the application is being tested. This provides detailed data on the behaviour of the OTS component, thus going beyond the mere collection

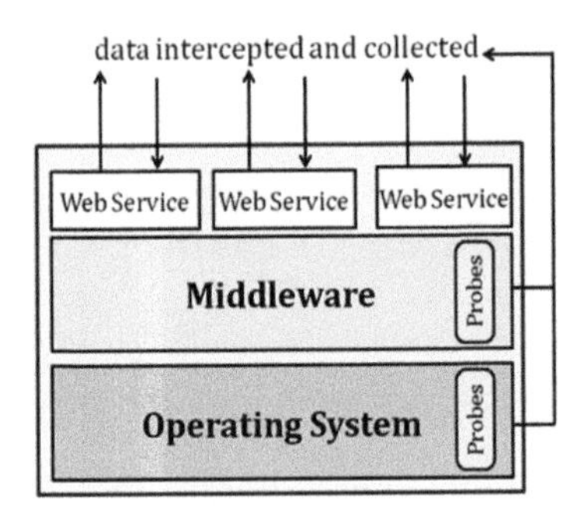

Figure 10.2 Detailed functioning of the Instrumented System.

of inputs and outputs or the monitoring of specific functions of the underlying system that the OTS component uses. To achieve this, it is necessary to instrument the kernel of the OS introducing monitoring probes that report the value of the selected variables per unit of time. These values should be stored in a standard format to later be externalized through the *Dataflow Out (DO)*.

As example of middleware, the environment may include an *application server* where the user can run the web applications and services that are necessary to be tested. Also, the application server includes monitoring of the values of relevant variables that are also stored in a standard format for later use of the *DO*. Due to the emerging role that web applications and services have in critical systems, the inclusion of a monitored application server is a very important requirement, as this allows gathering the values of variables that are closer to the applications under test.

The *Dataflow In (DI)* is necessary to perform the test in the applications. In the case of web applications and services, which have an interface available over the network, the DI In is constituted by the ports used to perform the tests together with OTS components that can execute the tests through these ports. The environment should also be ready to support the testing of other applications, with the Dataflow having the responsibility of translating the tests created by the testing tools in a form that can be executed in the target application. In practice, the Dataflow In represents the only part of Instrumented System that the user should implement in order to have his application tested.

10.2.2 Test and Collect

Test and Collect includes a set of extensible tools that should support the user in two main activities: (i) *testing*, and (ii) *storage and analysis*. The **testing** component controls the execution of the testing tools. Although the framework is designed to be fully automated, the human interaction cannot be completely avoided at least for the test execution.

The level of human interaction can vary from test to test; thus, each testing tool should provide its own interaction interface. It is mandatory that the testing tool communicates with the **storage and analysis** module to trace testing activity, providing information as test input/output, and executions results and durations that should be logged by the storage module to match the results provided by the IS during the execution of these tests. Figure 10.3 depicts this relationship, which is detailed below.

The **storage and analysis** module is also in charge of harvesting data from the IS probes and of structuring and storing them to facilitate the subsequent data analysis. The storage component is made up of three modules: (i) *Probes Collector* (PC), (ii) *Data Manager* (DM), and (iii) *Database* (DB).

The PC is responsible of reading data from IS probes and due to the different sources (middleware or OS) it needs to use different policies to respect the data availability and probe servers' constraints. Data read are then managed by the DM component that organizes the data coming from middleware structuring it to provide the state of the monitored system from a specific point of view.

Finally, data is stored in the underlying database. To provide more efficient data analysis capabilities, the template schema follows the model of a star schema from OLAP. In fact, a well-structured data repository and OLAP analysis can be very useful for analysis of results from dependability evaluation experiments [13]. Additionally, it makes possible to share and compare the results of multiple different experimental evaluations [13].

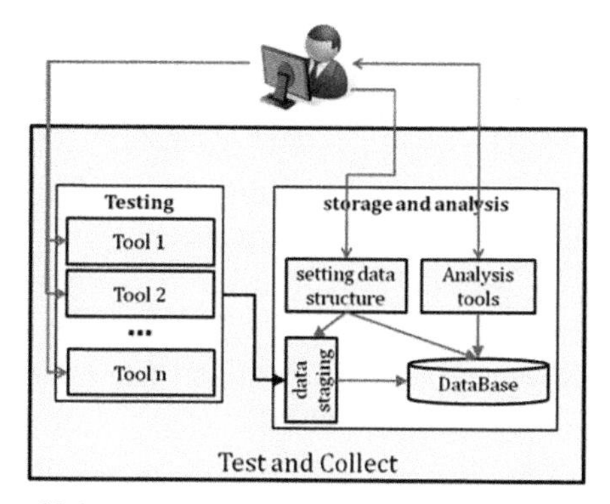

Figure 10.3 Detailed functioning of the Test and Collect.

The **testing tools** included should be able to generate different types of testing including functional testing, stress testing, robustness testing, penetration testing, security testing, etc. The definition of tests is always dependent on the type of application as well as specific to the domain of the application (e.g., testing requirements from standards). In the case of web applications and services, where the interfaces are usually well defined, the test generating tools usually require only minor configuration. However, in the case of other applications the user may be requested to configure or even extend the testing tools. To cope with this, the tools included are prepared to be easily extensible to accommodate the user needs. As the tools are easy to modify or replace, the framework provides high flexibility and makes it easier to test applications.

Functional testing is a black box testing technique that tries to find discrepancies between the program and the external specification [14] and it is based on a set of test cases derived from the analysis of the specification. Stress testing subjects the program to heavy loads or stresses [14]. In this case, the testing application must submit loads that match (or even surpass) the load that the application under test is specified to sustain over a period of time. This is particularly useful in web-based applications where you want to ensure that your application can handle a specific volume of concurrent users or requests. Robustness testing is a specific form of black-box testing. The goal is to characterize the behaviour of a system in presence of erroneous input conditions. Robustness testing stimulates the system in a way that triggers internal errors, exposing programming and design errors both in the error detection and recovery mechanisms. Penetration testing is a specialization of robustness testing that consists of the analysis of the program execution in the presence of malicious inputs, searching for potential vulnerabilities. Penetration testing tools provide an automatic way to search for vulnerabilities avoiding the repetitive and tedious task of doing hundreds or even thousands of tests by hand for each vulnerability type.

Finally, the toolset should be allowed to be used several data analysis algorithms; including fault detection mechanisms based in static threshold analysis algorithms and also statistical analysis algorithms. However, the main idea is to leave to the user the conditions to perform the analysis using the algorithms that he is more experienced with and, above of all, that are most adequate to his business domain. In fact, one strength of the use OLAP analysis techniques is the optimization of their schema for the use of *ad-hoc* queries [13].

10.3 Implementation Details

To show the applicability of the approach and perform an experimental evaluation, a prototype was designed and is currently under development. For cost reduction and to allow bigger flexibility, effort was made to use low licensing cost solutions resulting many times in preference for free or open source software. It is important that, in many cases, the selection of one technology to use impacts the technologies for other layers. The next sections detail the status of implementation of each one of the components of the framework as well as the technologies selected to use.

10.3.1 Instrumented System (IS) Implementation

The node component was implemented in a virtual machine. This option for virtualization provides flexibility as can be easily replicated and maintained. This will allow the deployment of the node ready to use in any number necessary for the system. The operating system selected to implement the prototype was CentOS [15]. First we narrowed our options to Linux based distributions due to the cost advantage and to the diversity of monitoring options to monitor the kernel events. From the multitude of options available, CentOS provides a free enterprise class OS.

The instrumentation of the operating system was implemented in the form of a loadable kernel module using the SystemTap tool [16]. This tool builds on and extends the capabilities of the *kprobes* [17] kernel debugging infrastructure and allows to program breakpoint handlers using a high-level scripting language that is later translated into C code. This way, SystemTap simplifies the development of system instrumentation and also improves the reuse of existing instrumentations, thus allowing building up on the expertise of others. The developers of SystemTap also took into consideration the portability and safety concerns, both of which have major importance in this work.

The prototype implementation of Instrumented System includes, as example of middleware, an application server with monitoring capabilities to allow testing the web applications and services. The selection of choice was JBoss Application Server (JBoss AS) [18], an application server that implements the Java Platform, Enterprise Edition (Java EE). It is free and open source software available under the terms of the GNU LGPL and it is written in Java and as such is cross-platform: usable on any operating system that supports Java. JBoss represents the industry *de facto* standard for deploying Java-based Web applications, it has a wide community acceptance, and support subscriptions can be purchased. For monitoring the values of the applications running inside the app server, it is used Java Management Extensions (JMX) Technology. JMX provides the tools for building distributed, modular and dynamic solutions for monitoring devices or applications. It is designed to provide high flexibility both for legacy systems and for the future. JMX is supported by the most relevant Java application servers. This will allow in the future, porting the monitoring solution to other application servers, and it was one of the requirements as it is planned to add other servers to the prototype to provide a broader range of options and compatibility to the user.

The implementation of the *Dataflow In* depends greatly on the applications to be tested. For the case of web applications and web services, OTS components together with the ports that allow the network traffic constitute the *Dataflow In*. For instance, in the case of web services, the toolset includes the open source tool *soapui* [19], that is the visible part of the Dataflow In for the user. This tool allows to easily executing user-defined tests in the web services under test. In other cases, where the test execution such be from inside the testing machine, it is under development a daemon to run in background accepting communication through TPC sockets and performing the required tests.

Finally, the dataflow out was implemented as folder where the monitoring systems can write the files and from where the Test and Collect can pool the files periodically. The data is split in chunks, each file containing the data respecting to a certain period of time that is identified in the filename. There are many ways to extract the files from the exterior, but the solution currently adopted consists of using *secure copy* (scp), is a protocol to securely transfer files between two hosts, based on the *SSH* protocol. This is a preliminary implementation that will be enough for experimental evaluation but as a more automated solution is under development and should replace it in a near future.

10.3.2 Test and Collect Implementation

The Test and Collect includes a set of testing tools ready to use. Most of these tools are black-box tools, and as the name shows, these tools view the program as a black box and are completely unconcerned about its internal behaviour. Our framework by analysing the values of the monitored variables uses information about the behaviour of the application in a transparent fashion to the testing tools. Most of the tools included also target web services, one of the main targets of the framework, as they are increasingly used in business-critical systems. They provide always a well-defined interface, allowing an easy use of automated tools. Other types of tools and also targeting other will be added in future versions of the framework. In very specific domains, the user will need to write the necessary tests and implement the necessary tools to use them.

The testing tools currently available allow performing functional and stress test, penetration test and robustness test. The testing framework has been developed minimizing the human interaction especially during the testing activities. With the present tools, the human interaction is indeed focused in the configuration phase, which must be performed one time for each tool. Such tools can provide common configurations as well as they can propose a configuration that suits the testing needs.

10.3.2.1 Functional and stress testing

Functional test is a quality assurance process based on black-box approach that aims to provide a proof of implementation correctness regarding the specifications of the software under test. The test is performed feeding the software under test with well-known values and examining the output produced.

Although common functional tests involve the test of single methods, within this context, it has been followed an approach which tackles high-level functionalities. The approach consists of a set of workflows that mimic the behaviour of a software user for executing specific high-level tasks, which in turn can comprise the invocation of a huge variety of methods [20].

The workflows definition is a cornerstone of this approach and it must be defined specifically for each service under test considering its interface and the software specification. Workflows define how and when the service interface of the software under test should be questioned and, also, they provide the information needed for the subsequent phase of result validation. Following the black-box approach the verification is done invoking specific methods of the service for checking its internal status.

The importance of these workflows is further emphasized since they can be used as bricks for compound and complex workflows for Stress testing. Stress testing is a form of deliberately intense testing used to determine the stability of a given system. The tool developed for functional test, properly configured with suitable workflows, can stimulate the system under test to provide evidence of stability. Workflows for Stress testing have been defined from the high-level tasks identified for the functional tests by parallelizing multiple high-level tasks invoked from a variety of users and abbreviating to the minimum the delay between sequential invocations.

10.3.2.2 Robustness testing and penetration testing

Robustness testing is a specific form of black-box testing that attempts to characterize the behaviour of a system in the presence of erroneous or unexpected input conditions [21]. The tool instrumented in the testing framework implements the technique proposed in [22]. The approach consists of a set of robustness tests that is applied during execution to disclose both programming and design problems.

The set of robustness tests is automatically generated by applying a set of predefined rules (see detailed list in [22]) to the parameters of each operation of the web service during the workload execution. An important aspect is that rules focus difficult input validation aspects, such as: null and empty values, valid values with special characteristics, invalid values with special characteristics, maximum and minimum valid values in the domain, values exceeding the maximum and minimum valid values in the domain, and values that cause data type overflow. The robustness of the web services is characterized according to the failure modes adapted from the CRASH scale.

Penetration testing is nowadays one of the most used techniques by web developers to detect vulnerabilities in their applications and services. This technique assumes particular relevance in the web services environment, as many times clients and providers need to test services without having access to the source code (e.g. when testing third-party services), which prevents the use of more effective techniques that require that access. The tool instrumented in the testing framework implements a technique targeting the detection of SQL Injection vulnerabilities in web services. The tool was originally presented in [11].

10.3.2.3 Data storage and analysis tools

A tool is under development for the generation of the star schema according to the needs of the user. This tool, based on the template star schema provided, and after some configuration by the user, generates the schema that will store the monitored data. This tool will also include capabilities to perform the extraction, transformation and load (ETL) of the data. It will allow the user to retrieve the data from the Instrumented Systems using the *Dataflow Out* channel and insert the data in the schema. With a wide range of options for ETL software available, the option for developing a new tool comes from simplicity reasons: it would be an exaggeration to use a heavy ETL tool while our toolset only needs for a very specific and simple tool that designed to work based on a the configuration that the user provides when creating the star schema for the DBMS. Also, the use of third party ETL tools would most probably require the user to have knowledge on how to operate them.

A myriad of solutions are available to implement the data storage. PostgreSQL[1] is an open source solution with a long history of development a proven architecture with recognized reputation for reliability, data integrity, and correctness. It gives to the framework great flexibility as it runs on all major operating systems, including Linux, UNIX, and Windows. A lot of tools from the community support PostgreSQL, and it is also widely supported by open source business intelligence tools as SpagoBI[2] and Pentaho community edition[3].

As aforementioned, in terms of data analysis, the main goal is to provide the user the best conditions for the execution of the analysis of his preference. With this goal, the tool set includes basis tools for data visualization and query execution. The toolset will also include the more advanced tools as the mentioned BI tools (SpagoBI and Pentaho CE) as well as other options that are also considered to be included in the toolset [23]. For better analysis, it is necessary that the data is correlated to the tests executed, and this is a very important part of the ETL process. Finally, the toolset will include ready to use tools that use some more specific algorithms targeting fault detection, proposed by research community. Examples of these works are static threshold analysis [9] and statistical analysis algorithms for on-line fault detection [8].

10.4 Demonstration

Two case studies were devised to demonstrate the applicability and feasibility of the approach.

The first case study is presented in Section 10.4.1 and uses the Life ray Portal [24], which is an enterprise web platform project that aims for immediate delivery of robust business solutions for organizations. An API based on SOAP web services is provided, containing a diverse range of functionalities. These services are an interesting case for testing the framework, after the framework is deployed on top of the instrumented JBoss AS middleware.

[1]www.postgresql.org
[2]www.spagobi.org
[3]http://community.pentaho.com

The second case study, presented in Section 10.4.2, is based on the PMF simulator. SHAPE is a system installed along a specific railway line. The main purpose of the system is to automatically detect anomalous and hazardous situations on the trains running on that line. SHAPE aims at detecting two specific situations: i) SHAPE can detect fires on board a train, through reading at a distance of the temperature of the external surface of the trains; ii) it is able to detect possible violations of the reference shape, through specific laser scanners, in order to identify any dangerous protruding part of the train.

10.4.1 Case Study: Life Ray Web Services

Life ray is free and open sourced Java software that was initially developed to provide an open source enterprise quality portal. Since the early stages of development, Life ray has been widely adopted for intranet as well as extranet enterprise solution. Eventually, it brought Life ray to have a big supporting community, which, together with the Life ray foundation, contributed to define a generic and extendible product.

Life ray Portal is an enterprise web platform project that aims for immediate delivery of robust business solutions for organizations. It includes features that are usually necessary for the development of websites and portals, as built-in web content and document management system. Life ray is developed using Java technologies and it is ready to work in a large set of web/application servers. In fact, the community edition is free and open source software available under Licensed under the terms of the GNU LGPL. It follows an extendible architecture by plugins that, in turn, encompass collaboration, social networking, and single sign on, per-component privileges policy as well as e-commerce tools. Third-party plugins are also available to provide more advanced feature such as Microsoft office integration. A plugin can be seen as a J2EE-Servlet and is referred to as a portlet. Portlets communicate with each other using the services that each one exposes that, in turn perform the portlets business logic. Our installation of Life ray includes the version 6.0 of the portal and 83 deployed SOAP (Simple Object Access Protocol) web services. More details on Life ray can be found in [24].

10.4.1.1 Tests performed

During this case study, different types of tests were applied: Functional, Stress, Penetration and Robustness tests. To verify the correctness of Liferay services, a study on its plugins interaction has been conducted. The necessity for the preliminary study has been felt because of the strictly correlated invocations among methods exposed by web services.

The preliminary study has been exploited to define workloads that could mimic the behavior of Life ray internal interactions, which correspond to **functional tests**. Even simple activity, like posting a message on the blog by UI interface, could involve a plethora of plugins including authentication, user information retrieving, permission checking and finally messaging service. To stimulate Life ray in a way that could resemble human activities, many Workloads have been defined to cover Life ray functionalities by mimicking the behavior of human interaction. Mimicked actions encompass posting a message in the blog and in the forum, creating an event in the calendar, creating a directory-tree in the file repository and uploading a file in it.

These workloads have been used to define other workloads for **stress tests,** to highlight possible weakness in terms of concurrency management. Those tests have been designed from the workloads defined for functional tests to evaluate Life ray behavior under a heavy load. For each workload, that mimics a specific action, a new one is defined as a composition of many copies of the same workload; these copies differ just for the user. The purpose of this approach is to simulate multiple users' activities on Life ray, that stimulate the same services and some shared data. The stress test, as it is designed, suits especially well when there isn't a sound knowledge of web services internal mechanism; the deeper is the knowledge of web service internals the more effective workloads can be designed.

Regarding **robustness testing**, due to the preliminary study performed for the functional test, a generic knowledge of methods invocation was available to configure the tool to generate better tests. This knowledge was especially important for the tool to use values that exercise the code of the web service under test in a more complete way. After the configuration of the tool it submits the robustness tests in an automated way and reports the robustness problems found.

As for **penetration testing**, the available knowledge of methods invocation was a key to configure the tool to generate better workloads and attack loads. This is especially relevant for this tool as its effectiveness is depending on the completeness of its workloads. After this configuration, the test execution is a straightforward process in which the tool submits its workload and attack load to the web service and then reports the vulnerabilities found.

10.4.1.2 Tests results

Experiment execution is made up of three phases, where just the final one is specific for the kind of test to be conducted. The phases are:

1. Set up Service Under Test (SUT),
2. Data Logger execution,
3. Testing tool execution.

On the first phase the service under test is started up. This phase includes also the startup of middleware and OS probes. The second phase can be launched simultaneously, as it does not read information from OS or middleware probes: it just prepares the structures needed for logging. During test execution, the testing framework logs raw tests results and prints on the console information on the tests execution (test currently running, test duration, etc.). This information is useful to monitoring the tests execution. Test results are collected during test execution; at the tests termination, collected data are flushed into a database.

Different tools that range from very specific tools such as R or MatLab to commonly available and general-purpose tool such as OpenOffice-Calc are installed on the Test and Collect system, connected to the database, and can be used to retrieve and analyze data.

Due to the wide usage and the extended support that Life ray received from its community since its development started, it was expected that Life ray passed all the functional tests defined.

Figure 10.4 shows an extract of the workload (set of services invocations) used for creating a new event on a calendar (it includes logging in to the system, listing the available/subscribed calendars, choose to first one and listing all the events of February, adding the new item then logging out; subsequently further invocations checks that the event was correctly recorded). The invocation correctness is verified by a visual inspection of Life ray services.

The output produced during the execution of the test is displayed on the screen of the Test and Collect system and consists of a sequence of services methods invocations. For each one of these, the HTTP response code is printed out. Being a functional test is mandatory that the HTTP response code would match with the expected one.

The aim of this **stress testing** is to assess the ability to resist against a workload which leverage on high frequency of requests, and the results can be evaluated in term of system loads and resource usage. Table 10.1 shows the average CPU usage and memory usage; the former one is furthermore

```
- <tns:Workload name="addEventInCalendar">
  - <tns:Choreography>
    - <tns:Call portlet="Portlet_Cal_CalEventService" method="getEventsCount/1">
      - <tns:Parameters>
          <tns:Parameter name="groupId" variable="groupId"/>
          <tns:Parameter name="start">0</tns:Parameter>
          <tns:Parameter name="end">1393592163</tns:Parameter>
        </tns:Parameters>
      </tns:Call>
    - <tns:Call portlet="Portlet_Cal_CalEventService" method="getEvents/1">
      - <tns:Parameters>
          <tns:Parameter name="groupId" variable="groupId"/>
          <tns:Parameter name="start">0</tns:Parameter>
          <tns:Parameter name="end">1393592163</tns:Parameter>
        </tns:Parameters>
      </tns:Call>
    - <tns:Call portlet="Portlet_Cal_CalEventService" method="addEvent/1">
      - <tns:Parameters>
          <tns:Parameter name="title">titoloEvento</tns:Parameter>
```

Figure 10.4 An extract of the workload to set a New Calendar Event.

Table 10.1 Extract test results for New Calendar Event

Parallel Requests (n_r)	Process CPU Load (%)	System CPU Load (%)	System Load Average	Free Heap Memory (B)	Free Non Heap Mem (B)	IO Written/ Read Data (B)
5	0.3031	0.38245	1.54	100128920	24899588	5959
10	0.1254	0.651	1.195	86411094	22482534	77344
100	0.1342	0.999	2.34	84833658	21993838	184244

detailed distinguishing between process CPU load and system CPU load, with system CPU load that encompasses any task running on the system. Memory usage is furthermore detailed as well distinguishing between heap memory, used for java objects and non-heap memory.

The table encompasses the experiments of 5, 10 and 100 simultaneous execution of the "New Calendar Event" workload. Data are collected 1 times per second. Table 10.1 shows that the CPU usage of service process remains quite stable despite the increase of the number of requests. Process CPU load, the system load as well as memory usage vary as the number of parallel requests increase. The table shows that system resources usage clearly increases due to the waits for Disk Output activities, which rise.

Figure 10.5 shows an extract of the **robustness test** report, in which all the tests reported robustness problems. This would suggest weakness in the services, but a manual inspection revealed that while tool reports "PROBLEM", the service correctly identify and discard the invalid request. We explain this with the help of Figure 10.6.

fact_robustness_test_result_id	field	type	code
1	Field{questionId} of LONG @ [0-1001[	ROBUSTNESS	PROBLEM
2	Field{questionId} of LONG @ [0-1001[	ROBUSTNESS	PROBLEM
3	Field{questionId} of LONG @ [0-1001[	ROBUSTNESS	PROBLEM
4	Field{questionId} of LONG @ [0-1001[	ROBUSTNESS	PROBLEM
5	Field{questionId} of LONG @ [0-1001[	ROBUSTNESS	PROBLEM
6	Field{questionId} of LONG @ [0-1001[	ROBUSTNESS	PROBLEM

Figure 10.5 Extract from robustness test results.


```
- <soapenv:Envelope>
    <soapenv:Header/>
  - <soapenv:Body>
    - <urn:addQuestion soapenv:encodingStyle="http://schemas.xmlsoap.org/soap/encoding/">
        <titleMapLanguageIds xsi:type="urn:ArrayOf_xsd_string" soapenc:arrayType="soapenc:string[]"/>
        <titleMapValues xsi:type="urn:ArrayOf_xsd_string" soapenc:arrayType="soapenc:string[]"/>
        <descriptionMapLanguageIds xsi:type="urn:ArrayOf_xsd_string" soapenc:arrayType="soapenc:string[]"/>
        <descriptionMapValues xsi:type="urn:ArrayOf_xsd_string" soapenc:arrayType="soapenc:string[]"/>
        <expirationDateMonth xsi:type="xsd:int">invalidNumber</expirationDateMonth>
        <expirationDateDay xsi:type="xsd:int">6</expirationDateDay>
        <expirationDateYear xsi:type="xsd:int">2000</expirationDateYear>
        <expirationDateHour xsi:type="xsd:int">5</expirationDateHour>
        <expirationDateMinute xsi:type="xsd:int">5</expirationDateMinute>
        <neverExpire xsi:type="xsd:boolean">true</neverExpire>
        <choices xsi:type="urn:ArrayOf_tns2_PollsChoiceSoap" soapenc:arrayType="mod:PollsChoiceSoap[]"/>
        <serviceContext xsi:type="ser:ServiceContext"> </serviceContext>
      </urn:addQuestion>
    </soapenv:Body>
  </soapenv:Envelope>
```


(a)


```
- <soapenv:Envelope>
  - <soapenv:Body>
    - <soapenv:Fault>
        <faultcode>soapenv:Server.userException</faultcode>
      - <faultstring>
          java.lang.NumberFormatException: For input string: "invalidNumber"
        </faultstring>
      - <detail>
          <ns1:hostname>testingBOX</ns1:hostname>
        </detail>
      </soapenv:Fault>
    </soapenv:Body>
  </soapenv:Envelope>
```


(b)

Figure 10.6 Example of robustness test: (a) request; (b) response.

Figure 10.6(a) shows an extract of a robustness test involving the Poll Service, in particular the "add Question" method. Life ray, relying on Axis2 for parsing values, automatically manages the invalid value for the parameter "expiration DateMonth" rejecting the request and without passing it to the "actual" service. The rejection causes an HTTP 533 (which belongs to the "internal error" family): the tool used for robustness testing, operating at black-box, can't distinguished this answer from any other internal error, and consequently the "PROBLEM" code is displayed in Figure 10.6(b) which shows the response that Life ray produces for the request.

Life ray uses Axis2 for service publishing and interface, Axis2 is responsible for parsing values passed by SOAP as well as for invoking the actual Java method which was remotely requested. The parsing phase consists also of a validation phase in which the parsed values are validate against their destination types constraints. The failure of this phase implies the subsequent rejection of the request and thus the generation of a response with HTTP code 500.

Figure 10.7 shows an extract of the results of the **penetration tests** applied to Life ray Calendar Service. The extracted data, as well as the entire test results, show the robustness of Life ray against penetration attacks. All the potentially risky requests are identified and discarded by the Axis2 Layer for services interface, by the Object Relational Mapping (ORM) layer for objects persistency and by the permission checking mechanism, which constitute a cornerstone for Life ray services interoperability.

In fact, Life ray exposes its services using Axis2, which validates the invocation parameters before passing the request to the "actual" service. Additionally, Life ray relays upon Hibernate (the ORM used) which provides an SQL parameter sanitizing service, which in turn it uses named queries that work on top of statements of the JDBC API; all those layers operate the necessary actions to avoid risks from malicious requests. Finally, the invocations that include items the user is not authorized to use are identified by the Life ray Permission Service.

fact_penetration_test_result_id	field	type	code	fact_id
1	Field{questionId} of LONG @ [0-1001[	SQL	PASSED	56151
2	Field{questionId} of LONG @ [0-1001[	SQL	PASSED	56165
3	Field{questionId} of LONG @ [0-1001[	SQL	PASSED	56169
4	Field{questionId} of LONG @ [0-1001[	SQL	PASSED	56174
5	Field{questionId} of LONG @ [0-1001[	SQL	PASSED	56204

Figure 10.7 Calendar Service penetration tests result.

10.4.2 Case Study: SHAPE

We used a second use case to show the flexibility of the approach and also to demonstrate that the approach is not depending on the concrete technological implementation. This use case was based on SHAPE, which is a system installed along a specific railway line. SHAPE has the requirement that the all the tests to be performed must be done in an environment certified as equivalent to the target environment.

The main purpose of the system is to automatically detect anomalous and hazardous situations on the trains running on that line. In particular, SHAPE aims at detecting two specific situations: i) SHAPE is able to detect fires on board a train, through reading at a distance of the temperature of the external surface of the trains; ii) it is able to detect possible violations of the reference shape, through specific laser scanners, in order to identify any dangerous protruding part of the train.

SHAPE was designed to be suitable for interfacing with existing signalling systems, thus to send possible alarms useful to safely stop the train and to properly manage the critical detected event, according to the foreseen recovery actions. SHAPE is composed by the components: Scanner, Init & Diagnosis, Data Acquisition, Data Aggregator, Data Analyser and Monitor, System State.

- Init & Diagnosis – communicates with the scanner in order to collect diagnostic data and to trigger scanner activation.
- Data Acquisition – receives raw data from scanners.
- Data Aggregator – receives train data from Data Acquisition (e.g. images produced by the scanner) and aggregates such information, to be sent to the Monitor component.
- Data Analyser – receives aggregated data from the Data Aggregator and send analysis results to the Monitor component.
- Monitor – manages all the system states phases according to the data received from the Shape Component.
- System State – acquires information regarding the system state from each component and sends them to the WaySide component.

10.4.2.1 Monitoring environment adaptation

SHAPE has stringent requirements whichneeds to be tested in the same operating system, configuration, and equivalent hardware that it is supposed to be used in the future. Therefore, to be able to use the monitoring and testing approach in together with SHAPE, it was necessary to port the monitoring facilities to the target system and configuration.

The new system uses a different operating system and due to criticality restrictions, it cannot have new packages installed, as the system monitoring tools (SystemTAP) required by used in the implementation of the Instrumented System. Therefore, the challenge was to implement similar monitoring functionalities with less intrusive solutions.

The solution used the following tools, which are present in most of unix and linux distributions:

- top – provides data about cpu and memory usage;
- mpstat – provides data about system load;
- iostat – provides data about I/O usage;

Another tool was necessary because the SHAPE simulator involves a set of processes and subprocesses that are continuously evolving.

- pstree – allows to track the processes and the respective process tree, so it is possible to gather data about all the processes relevant for the monitoring system.

The downside of this solution is the performance. In practice, although less intrusive than the SystemTAP solution, it takes much more time to obtain data, and therefore it does not allow small gathering windows. However, we believe that the window is still small enough to do fine grained analysis of the system behavior.

10.4.2.2 Tests performed

To demonstrate the solution, we executed the SHAPE simulator during 24h while monitoring the relevant variables of the system. During this period, the simulator was exercised using the test cases available to test the correctness of his responses. At the same time, the newly included probes seamlessly monitored the variables of interest. Table 10.2 contains the summary of the most relevant variables monitored during the period.

All the variables were analyzed are stored for each sampling instance. This allows us to do temporal analysis of the variables. Figure 10.8 presents the evolution of one specific variable over time, in this case the "Number of SHAPE processes". As we can observe, the amount of CPU usage keeps increasing throughout the collection period, but still in relatively short values.

Table 10.2 Summary of the variables monitored

Variable	AVG	STDV	MIN	MAX
Total User CPU	0.14	0.08	0.00	0.30
Total System CPU	0.24	0.12	0.00	0.50
Average User CPU	0.14	0.07	0.02	0.26
Average System CPU	0.25	0.12	0.03	0.45
Average IO Wait CPU	0.11	0.00	0.11	0.12
Memory Used	837945	32356	782680	903032
Memory Free	1031399	32356	966312	1086664
Memory Cached	188067	10077	168972	205268
Swap Used	0.00	0.00	0.00	0.00
Swap Cached	582445	20468	547568	618732
IO Disk Read Per Sec	1.78	0.03	1.74	1.83
IO Disk Write Per Sec	7.15	0.22	6.50	7.52
IO Disk Read	2913344	30	2913256	2913424
IO Disk Write	11674444	527703	10347224	12589656
# of SHAPE Processes	2.98	0.17	1.00	4.00
Number of Samples	*151146*			

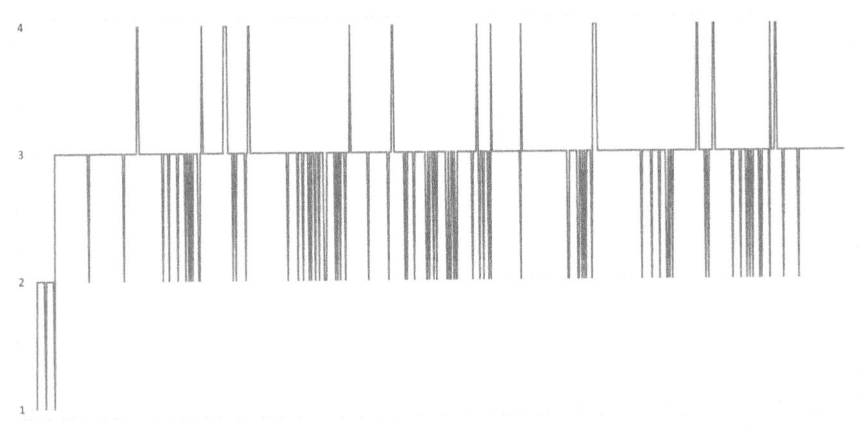

Figure 10.8 Evolution of Number of working processes in SHAPE.

10.5 Conclusion

In a context where OTS components are increasingly used on critical scenarios, companies need tools that help them to understand the quality of these components. In specific cases of testing, rather than using their own developed ad-hoc and poorly-reusable testing tools, these companies can benefit from using cost effective techniques and tools.

This chapter presented a reusable and adaptable framework for testing and monitoring of critical OTS applications and services that includes an instrumented box for monitoring OS and application level variables, a testing toolset that is adaptable for testing the target components, and tools for data storage and analysis. The architecture of the framework was described as well as the status of its implementation.

The framework allows users to easily apply functional testing, stress testing, robustness testing and penetration testing to their web services. The procedure to use the framework is described and its usability is illustrated with a case study that uses the Life ray platform, composed of several web services. The case study shows how flexible is the framework, allowing integration of multiple third party tools seamlessly. Obviously, in the case of functional testing, it is necessary to conduct some preliminary study to emulate its use cases, but this is expected due to the nature of the tests. The framework can orchestrate the use of the tools and reduce the human effort by reutilizing the information provided at configuration time within multiple tools.

The concepts behind the framework can also be extended to setups that differ from the ones defined in the framework implementation. This was demonstrated in the second use case, in which the concepts.

Future work includes the integration of failure detection and prediction algorithms in the box. Additionally, the framework can be modified to use more than one Instrumented System at the same time, allows testing more complex systems. Finally, it can be extended to take advantage of other kinds of information monitoring.

References

[1] Tran, E. (1999). "Verification/Validation/Certification," in: *Topics in Dependable Embedded Systems*, ed. P. Koopman. Carnegie Mellon University, Pittsburgh, PA.

[2] IEC. (1998). *IEC 61508 TC: IEC 61508, Functional Safety of Electrical/ Electronic/Programmable Electronic (E/E/PE) Safety Related Systems, Part 3: Software Requirements*. IEC, Geneva, Swiss (1998).

[3] RTCA. (2011). *RTCA: RTCA DO-178C/EUROCAE ED-12C – Software Considerations in Airborne Systems and Equipment Certification.*

[4] ISO. (2011). *ISO: Road vehicles – Functional safety – Part 6: Product development at the software level.*

[5] IEEE Computer Society. (2012). *Software & Systems Engineering Standards Committee: 1012–2012 – IEEE Standard for System and Software Verification and Validation.*

[6] Ghosh, A. K., Schmid, M., and Hill, F. (1999). "Wrapping Windows NT software for robustness. In: Fault-Tolerant Computing," in *Twenty-Ninth Annual International Symposium on Digest of Papers* (New York, NY: IEEE), 344–347.

[7] Popov, P., Strigini, L., Riddle, S., and Romanovsky, A. (2001). "Protective Wrapping of OTS components," in *Proc. 4th ICSE Workshop on Component-Based Software Engineering: Component Certification and System Prediction*, Toronto.

[8] Brancati, F. (2012). Adaptive and Safe Estimation of Different Sources of Uncertainty to Improve Dependability of Highly Dynamic Systems Through Online Monitoring Analysis (New York, NY: IEEE).

[9] Carrozza, G., Cinque, M., Cotroneo, D., and Natella, R. (2008). "Operating system support to detect application hangs," in *International Workshop on Verification and Evaluation of Computer and Communication Systems, VECoS*, Leeds, UK.

[10] Voas, J. M. (1998). Certifying off-the-shelf software components. *Computer* 31, 53–59.

[11] Antunes, N., and Vieira, M. (2009). "Detecting SQL Injection Vulnerabilities in Web Services," in *Fourth Latin-American Symposium on Dependable Computing (LADC '09)*, 17–24. IEEE Computer Society, Joao Pessoa, Brazil.

[12] Laranjeiro, N., Canelas, S., and Vieira, M. (2008). "wsrbench: An On-Line Tool for Robustness Benchmarking," in *IEEE International Conference on Services Computing, 2008* SCC '08 (New York, NY: IEEE), 187–194.

[13] Madeira, H., Costa, J., and Vieira, M. (2003). "The OLAP and data warehousing approaches for analysis and sharing of results from dependability evaluation experiments," in *Proc. of 2003 International Conference on Dependable Systems and Networks (DSN 2003)* (New York, NY: IEEE), 86–91.

[14] Myers, G. J., Sandler, C., and Badgett, T. (2011). *The art of software testing.* Hoboken, NJ: John Wiley & Sons.

[15] CentOS Project. *The Community ENTerprise Operating System.* Available at: http://www.centos.org/

[16] Prasad, V., Cohen, W., Eigler, F. C., Hunt, M., Keniston, J., and Chen, B. (2005). "Locating system problems using dynamic instrumentation," in *2005 Ottawa Linux Symposium* (New York, NY: IEEE), 49–64 (2005).

[17] Moore, R. J. (2001). "A Universal Dynamic Trace for Linux and Other Operating Systems," in *USENIX Annual Technical Conference, FREENIX Track*, Boston, MA, USA, 297–308.

[18] Red Hat. *JBoss Application Server.* Available at: https://www.jboss.org/jbossas/

[19] eviware: *soapUI.* Available at: http://www.soapui.org/

[20] Ceccarelli, A., Zoppi, T., Bondavalli, A., Duchi, F., and Vella, G. (2014). "A testbed for evaluating anomaly detection monitors through fault injection," in *5th IEEE Workshop on self-organizing real-time systems (SORT 2014)*, Reno, Nevada, USA.

[21] Koopman, P., and DeVale, J. (1999). "Comparing the robustness of POSIX operating systems," in *Twenty-Ninth Annual International Symposium on Fault-Tolerant Computing Digest of Papers* (New York, NY: IEEE), 30–37.

[22] Vieira, M., Laranjeiro, N., and Madeira, H. (2007). "Benchmarking the Robustness of Web Services," in *13th Pacific Rim International Symposium on Dependable Computing, 2007 PRDC 2007* (New York, NY: IEEE), 322–329.

[23] Golfarelli, M. (2009). "Open source BI platforms: a functional and architectural comparison," in *Data Warehousing and Knowledge Discovery* (Berlin: Springer), 287–297.

[24] Liferay, Inc. *Liferay Portal.* Available at: http://www.liferay.com/

11

Validating a Safety Critical Railway Application Using Fault Injection

Ivano Irrera[1], András Zentai[2], João Carlos Cunha[1,3] and Henrique Madeira[1]

[1]CISUC, Department of Informatics Engineering, University of Coimbra, Portugal
[2]Prolan Process Control Co., Szentendrei út 1-3, H-2011 Budakalász, Hungary
[3]ISEC – Coimbra Institute of Engineering, Polytechnic Institute of Coimbra, Portugal

The need for safety assurance in critical systems demand for new tools and techniques which are able to provide the required confidence while maintaining the costs relatively at a low level. Fault Injection (FI) is a technique extensively used in several domains, such as space, but sporadically used in the railways. In this chapter, we present a fault-injection tool able to complement the traditional verification and validation procedures, to validate the safety of ProSigma, a Safety Integrity Level (SIL) 4 safety-critical system for railway signaling, implementing a Triple Modular Redundancy (TMR) architecture. This tool is based on the Joint Test Action Group (JTAG) technology, and allows emulating the effects of hardware faults. Results from the FI campaigns show the ProSigma system exhibiting a high degree of tolerance to most of the injected faults, and unexpected behavior in some cases. The results also confirm the efficacy of the proposed technique to help understand worst-case scenarios for validating safety of such a critical system.

11.1 Introduction

The products in all technical and societal domains are required to be certified against hidden design and implementation defects that may induce malfunctioning, which may cause critical damage to the system itself, including the environment and humans. Several accidents caused by malfunctioning systems are sadly known, going along with the human technological rise.

A safe system is a system that will not cause harm to its users and the environment, in case a malfunctioning occurs. *Safety*, thus, comes to be an attribute of systems, which corresponds to guarantee the absence of catastrophic consequences on the user(s) and the environment to a certain extent. A system whose malfunctioning is likely to cause harm to users or the environment is named a *safety-critical system.*

Since their first realization, railway system fall into the class of safety-critical systems. For assuring the safety of such systems, best practices and standards have been proposed and used along with their technological evolution. In the last decades, a new series of standards have been proposed, namely the CENELEC standards (e.g., EN 50126, EN 50128, EN 50129, EN 50159), for regulating the development and safety assessment of software and hardware. In particular, the EN 50128 describes methods to be used in order to provide software that meet the demands for safety integrity. Although not directly referring fault-injection as a possible technique for verification and validation (V&V) processes, this is an approach extensively used in other domains, such as space.

Fault injection (FI) consists of the deliberate insertion of faults (i.e., realistic perturbations) in computer systems components in order to evaluate the dependability and safety properties of systems or to validate specific fault handling mechanisms. As typically FI tools perform FI campaigns with minimal user intervention (ideally, the process is fully automatic), it is possible to perform very large number of experiments (very often thousands or even millions), which makes FI a valuable method to anticipate worst-case scenarios or rare failure modes that are very hard to anticipate using analytical modeling or simulations techniques.

In this chapter, we present a FI tool and report some preliminary results meant to validate a TMR system for railway signaling. In the field of railway interlocking systems copper-based, long-distance connections exists between relay switches and remote equipment. In the case of constructing a new system, such as the one reported in this paper, the state-of-the-art solution is to apply Internet Protocol (IP) based signal transmission using Global System for Mobile communications (GSM) or fiber-optic communication. These new technologies pose significant safety challenges, which constitutes a relevant scenario for using FI.

The rest of the chapter is organized as followed: Section 11.2 presents background on V&V processes, certification and standards of railway systems, and FI; Section 11.3 presents the ProSigma system railway signaling system; the proposed FI tool-based on On-Chip Debugging (OCD) technology is described in Section 11.4. The experimental application of the proposed FI tool for the validation of the safety of the ProSigma signaling system is presented in Section 11.5, where results obtained are discussed. Section 11.6 concludes the chapter.

11.2 Fault Injection for V&V and Certification

Several approaches have been proposed to assess and guarantee the correct functioning of a given product. Among these systems V&V are activities that allow verifying whether a product meets its own requirements (Verification), and that the product does what is expected to be done (Validation). However, the application of V&V is challenging, as the definition of methods, strategies, and tools for verifying and validating a system adequately, while simultaneously keeping the cost and delivery time reasonably low, is inherently complex. Companies, in fact, are often, on one hand, pushed towards meeting predefined quality goals, and on the other hand, required to deliver systems at acceptable cost and time to market. It is not rare to find companies following a brute-force approach, by focusing large volume investments into tooling and in-house training, especially when coming down to the development of mission- and safety-critical systems.

Validation and verification are time-consuming activities in traditional software engineering even for non-critical applications. In the case of safety-critical systems, which are often embedded, the complexity of V&V and certification procedures are exacerbated by the need of keeping properties such as safety or availability, and by involving custom and Commercial Off-The-Shelf (COTS) hardware elements and application dependent-interfaces, resulting in an extremely large number of potential factors.

Safety-critical systems also required, over time, the creation of a field of study particularly aimed at focusing on safety-related issues: safety engineering. Safety engineering is a well-established field, including several techniques for the assurance and assessment of safety in a system. Among these, Failure Modes and Effects Analysis (FMEA), Preliminary Hazard Analysis (PHA), and Fault-Tree Analysis (FTA) are some of the most used techniques. In particular, FMEA is a technique that aims at collecting the known system's failure modes, and studying its propagation paths through the system and its effects. Failure Modes, Effects and Criticality Analysis (FMECA) is a version of FMEA in which criticality is taken into account, aiming to identify all critical and catastrophic subsystem or system failure modes.

FMEA is performed mainly manually, even though several works for an automated FMEA have been proposed [1].

Such techniques are expected to be part of the V&V process of a safety-critical system, and even be mandatory. In this direction, standards started to rise with the aim of reducing risks related to the use of safety-critical systems.

11.2.1 Standards for Safety-critical Railway Applications

Standards have been proposed for developing safety-critical systems, both general and domain specific and suggesting strategies, processes, and techniques to adopt along the entire development cycle.

The specification, design and validation of dependability-related aspects concerning *railway* applications are regulated by the CENELEC standards. The most important European standard concerning robustness in this field is the standard EN 50128:2011 – Railway applications – Communication, signaling, and processing systems – Software for railway control and protection systems (EN 50128) [2]. The EN 50128 gives indication about the lifecycle that has to be followed, the techniques and measures to be applied, the necessary competences, and the expected documents and their content. The Software Requirements Specification shall express the required properties of the software being developed. These properties, which are all (except safety) defined in ISO/IEC 9126 series, shall include (among others) robustness and maintainability. The Software Verification Plan shall address (among other properties) the evaluation of the safety and robustness requirements (defined in the Software Requirements Specification). Several techniques and methodologies are also indicated for ensuring the software robustness properties, as Software Error Effect Analysis (i.e., SW-FMEA).

Furthermore, the EN 50128 concentrates on methods that need to be used in order to provide software that meet the demands for safety integrity. The EN 50128 defines robustness as the "*ability of an item to detect and handle abnormal situations*". The most important of software techniques to assess and increase the robustness are the following: Defensive Programming, Information Encapsulation, Fault Detection and Diagnosis, Error Detecting and Correcting Codes, Diverse Programming, Software Error Effect Analysis, Control Flow Analysis, Common Cause Failure Analysis, FI, Boundary Value Analysis, and Coding Standard.

Generally, Railway Safety Cases shall provide evidences that the consideration of Robustness (error cases, abnormal inputs, etc.) is provided together with the system validation and verification.

According to the standard CENELEC EN 50129:2003 [15], safety-related software has been classified into five safety integrity levels, where 0 is the lowest and 4 is the highest. To be conforming to SIL 4 requirements, the safety availability of the equipment must be over 99.999%. From the safety functionality point of view (CENELEC EN 50129:2003 [15]), SIL is a number that indicates the required degree of confidence that a system will meet its specified safety functions with respect to systematic failures. From the software point of view, CENELEC EN 50128:2011 [14] defines software safety integrity level as a classification number that determines the techniques and measures that have to be applied to software.

11.2.2 Fault Injection

Fault injection is a technique consisting in deliberately injecting faults (e.g., bombarding devices with radiations) or modifying parts of the system in a way that emulates the presence of such faults.

Fault injection has been used extensively in research and also already recommended by several standards, such as space [3] and automotive [4] industry standards, in addition to Information and Communication Technology (ICT) industry in general [5]. The space industry, in particular, has a long tradition of using FI as part of the V&V activities, namely to simulate the effects of cosmic radiation in on-board systems. As mere examples, here are some references for the interested reader [6–8]. There are also some examples of the use of FI in the railway industry [9, 10].

Faults are the hypothesized cause of an error (an unexpected internal state of a system) that can lead to a system failure (e.g., crash, performance degradation, or any interruption of the service provided by the system) [11]. Hardware faults, such as bit-flip and stuck-at, occur in hardware components, while software faults are defects in a piece of software that exist due to some issue during the development phase, such as a missing system specification or poor testing. FI consists of deliberately inserting faults into a system in a way that emulates real faults [12]. It is a well-known approach used in many works, where the observation of systems in the presence of faults is needed, such as for fault tolerance and dependability validation [13, 14], estimation of fault-tolerance parameters [12], and benchmarking [15].

The type of faults injected typically fall into three kinds: hardware faults (e.g., bit flips), software faults (i.e., bugs), or input corruption at component interface level (often named as robustness testing). Although the initial FI tools are used to hardware approaches to inject (hardware) faults, including pin-level, heavy-ion radiation, and electromagnetic disturbances, modern FI tools use software approaches to inject the faults (actually, faults are emulated by software by mimicking the fault effects through the injection of errors). As modern FI tools use software to inject/emulate the faults, a key issue is the precision of the fault models. That is, the injected faults should be representative of the real faults that affect systems in the field. This is not a problem for the hardware faults, as the classic bit-flip or bit stuck-at models (at the processor register or memory level) are widely accepted, but the injection of realistic software faults (i.e., bugs) is far more complex. In software FI, the goal is to inject software faults (bugs) in a given software component to emulate the erroneous behavior that may result from the activation of residual bugs that may exist in that component. In this way it is possible to evaluate whether the system can cope up with the failures in the target software component or not, or to perform an experimental estimation of the risk of (re-) using software components.

An example of a survey of the earlier FI methods can be found in [16] and a very recent and extensive survey (57 pages) covering software FI is in [17], where the issue of defining realistic software fault models is explained in detail.

11.3 The ProSigma Safety-critical Railway Interlocking System

ProSigma [18] is a versatile Hardware–Software (HW–SW) system designed primarily for railway trackside signaling and communication purposes. It is a Safety Signal Transmitter (SST), which provides fail-safe signal transmission with high availability. It captures the analog signal outputs of the railway interlocking system, processes and transmits this information to a remote control center (DaKo). The ProSigma system is designed to be SIL 4 certified according to CENELEC EN 50126-1, 50126-2, 50128, and 50129 standards [12–15].

In case of disconnection or system failure, the outputs move into a safety position. The system is built from modular cards installed in racks, which enables system designers to scale the system according to the application needs.

11.3.1 Concepts of Generic Product, Generic Application and Specific Application

To ease the certification process, the system software is designed to have a three-layered architecture as it can be seen in Figure 11.1.

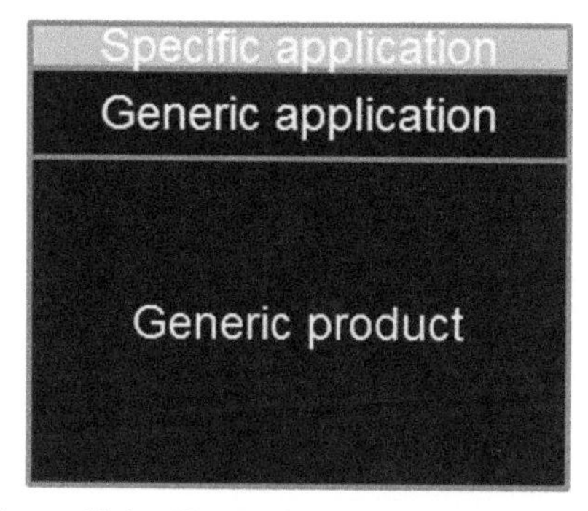

Figure 11.1 The ProSigma abstraction layers.

The bottom layer, called Generic Product (GP), implements the common functionalities of the system, including time synchronization, handling Controller Area Network (CAN) communication and other HW interfaces. The GP is quite complex, but it has to be certified only once, as it is common to all applications.

The middle layer, called Generic Application (GA), is a lightweight software component running on the top of the GP. Each GA handles one railway object (e.g. railway traffic signal, switch, etc.). Because of the simplicity of the code, the certification process of GA is relatively easy.

In the deployment phase of the system the GAs has to be parameterized with the actual values of the specific environment (e.g., voltage comparator thresholds, sampling frequency, etc.), which result in Specific Applications (SAs), which are the top layer of the software architecture. In the ProSigma system, the Logic and Input (LI) cards implement these three-layers design architecture.

11.3.2 The System Architecture and Functionality

A ProSigma test system was built in a pilot project to assess the benefits and drawbacks of FI, whose experimental results are presented in this chapter. The system has identical functionality but limited number of components compared to the one which is deployed trackside. The system adds a networking layer on top of a conventional relay based interlocking system. This network layer transmits the railway object states – represented by the relay outputs – to a remote control center (DaKo). The system architecture (Figure 11.2) consists of the following components:

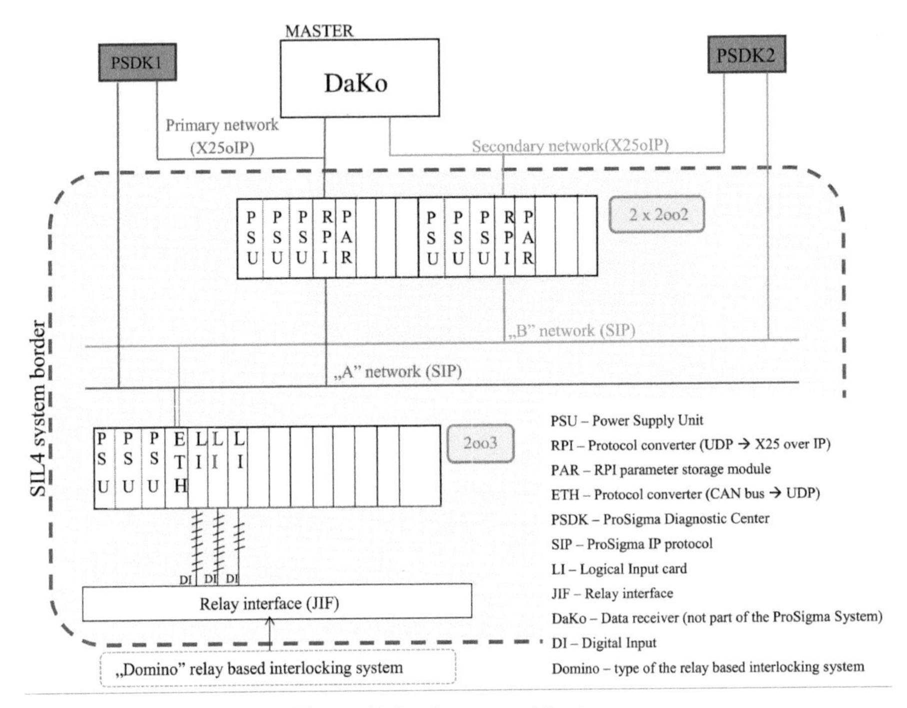

Figure 11.2 System architecture.

- Power Supply Unit (PSU), which supply 3.3 and 24 V of DC voltage to the cards;
- An analog signal conditioning unit (JIF) which filters and down-scales the relay output voltages from 0–48 V to 0–3 V range.
- Logic and Input card (LI) which are sampling the input voltages. They also contain the railway logic.
- CAN to UDP protocol converter cards (ETH), which convert CAN messages to UDP packets.
- UDP to X25 over IP protocol converter cards (RPI), which convert UDP packets to X25 over IP telegrams.
- Two diagnostic centers, which are responsible to log status and communication information and to provide diagnostic data to the operator.

11.3.2.1 Logic and Input (LI) card

The input signals of the system are the analog output voltage signals of a relay-based railway interlocking system: "Domino 70". These voltage signals are passed through a relay interface unit (JIF), which performs the voltage level interfacing for the Digital Inputs (DI) of the LI card. The Logic and Input (LI) card is a TMR system composed of three microcontrollers from different controller families. Red, Green and Blue (R, G, B) are the codenames for the three channels.

The Logic and Input (LI) cards are reading the analogous input signals and interpret them according to the rules of the specific railway object, which they are connected to. Finally LI cards transmit the status of the railway object states via CAN bus. The three channels (R, G, B) communicate on separate CAN buses, which are located on the back-panel of the mounting rack. See Figure 11.3 for the LI card.

The firmware (FW) of the controllers has been developed by different SW teams to avoid common mode faults. LI cards follow the three layered SW architecture described before consisting of two different FWs: GP and GA are parameters for the SA. On each channel, the FW of GP and GA are running on the microcontroller in a time and space partitioning architecture. On all channels, FW of the GPs handle the A/D conversion of the input signals. The raw data of the converted signals are filtered with a SW implemented de-bouncing algorithm in the GP to filter out the high frequency glitches of the relays. The GP FW calls the GA FW every 32 ms and the de-bounced values of the input signals are passed to the GA. The GA implements the railway object.

The railway object used in this case study is called block direction, which contains the information of the direction of traffic on the actual railway

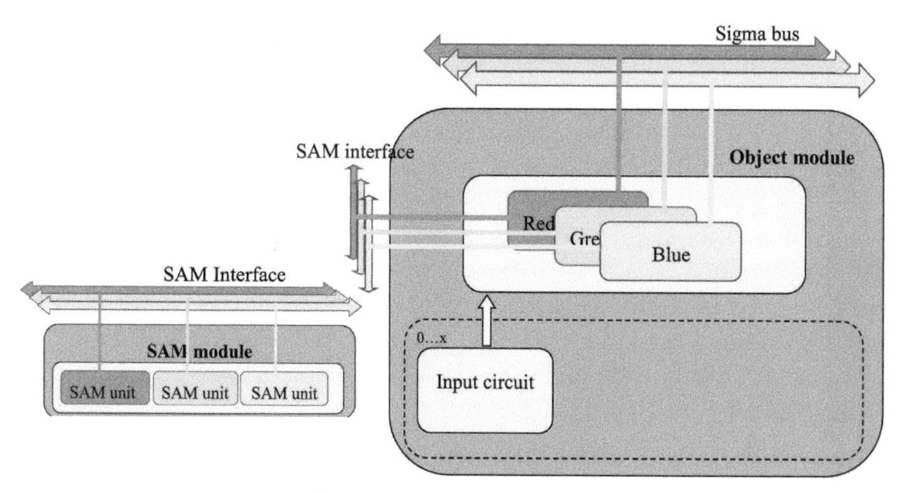

Figure 11.3 LI card interfaces.

Table 11.1 Railway object outputs

Valent Input	Antivalent Input	Meaning in Case P1 = 0	Meaning in Case P1 = 1
0	0	Transient (0 × 20) or invalid (0 × 80) state	Transient (0 × 20) or invalid (0 × 80) state
0	1	Direction = Exit (0 × 02)	Direction = Entry (0 × 01)
1	0	Direction = Entry (0 × 01)	Direction = Exit (0 × 02)
1	1	Transient (0 × 20) or invalid (0 × 80) state	Transient (0 × 20) or invalid (0 × 80) state

segment. The object has one input encoded by a pair of valent-antivalent input signals. Depending on the input signals and the value of parameter P1 the meaning of the direction could be entry, exit, transient or invalid as it is described in Table 11.1. The valent-antivalent signal pair does not change simultaneously so for a short period of time invalid input patterns (00 or 11) are accepted as transients. After that time period is passed, the signals became invalid.

The interpreted railway object state, encoded in the hexadecimal numbers indicated in Table 11.1, is transmitted on the CAN bus. The Sigma bus in Figure 11.3 indicates a proprietary application layer protocol implemented on top of the CAN bus. Specific Application Module (SAM) contains the parameters for the Generic Application. In the SAM module, 3 Flash memory chips contain the parameters for the three channels R, G, B. The LI card reads the parameter values from the memory via Serial Peripheral Interface (SPI) bus.

Interfaces of a LI card can be seen in Figure 11.3.

Up to 10 railway object modules could be inserted in one rack. In case there are more than 10 railway objects in a system, then the extra object modules are inserted into multiple racks. The racks are connected together to form a Local Area Network (LAN) using ETH cards.

11.3.2.2 ETH card

Primary function of the CAN to UDP protocol converter (ETH) card is to collect the railway object state information of the three channels from the CAN bus and transmit these messages as UDP datagrams on the Ethernet network. As it can be seen from Figure 11.4, ETH cards are connected to the CAN buses of all the three channels of the LI cards. This connection is physically realized through the back panel of the modular racks. Each ETH card contains two identical HWs. The inputs from both HWs are the

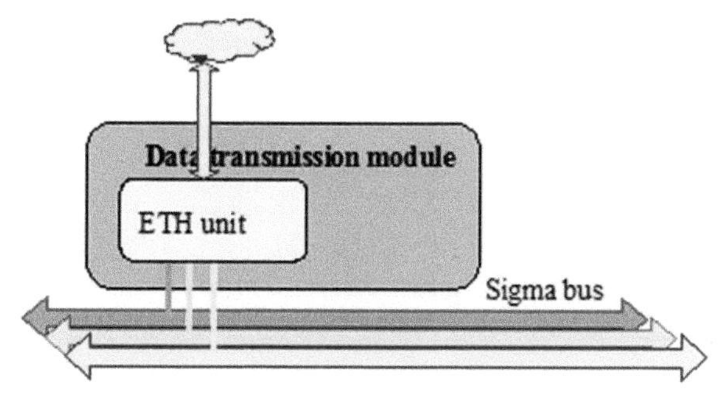

Figure 11.4 ETH card architecture.

same CAN channels, while the outputs are connected to two distinct LAN networks.

11.3.2.3 RPI card

The UDP messages are transmitted to the UDP to X25 over IP protocol converter unit (RPI). RPI architecture is depicted in Figure 11.5. This unit is responsible for converting the UDP packages to X25 over IP telegrams and sending these to the data receiver (DaKo), which is not part of the system.

The RPI module also performs a voting on the data collected from the three channels (extracted from the UDP packets), thus being central to the correct functioning of the TMR schema. Moreover, it provides two times 2-out-of-2 fault tolerance schema applied to both received data and voting result: the information is analyzed from two separated nodes (here named node 0 and node 1), and differences among data cause the entire RPI node to fail. Each node has a 2-out-of-2 architecture.

The underlying hardware of the UDP to X25 over IP protocol converter card/RBC-Prolan Interface (RPI) card is identical to the ETH card.

The functionality of the RPI card includes:

- Managing X25 connection with the Radio Block Center;
- Voting about the object states;
- Transmitting object states to the RBC;
- Exchanging Heartbeat (HB) signals both on active and on the potentially active channel.

11.3.2.4 Power Supply Units

In each rack, three Power SUpply (PSU) cards provide the necessary energy for the operation of the system.

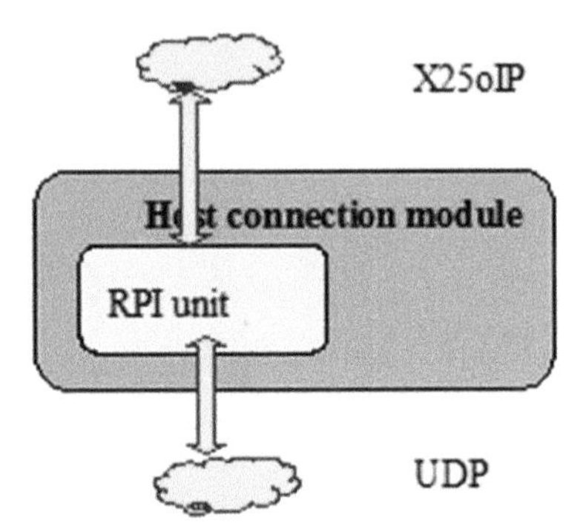

Figure 11.5 RPI card architecture.

11.3.2.5 Diagnostic centers

Two diagnostic centres (PSDK1 and PSDK2) are monitoring and logging the traffic on the internal and external networks.

11.3.2.6 Parameter modules

The parameter modules (PAR) contain the parameters of the GP and GA, which are required for the operation of the system.

11.3.3 System's Critical Aspects Worth to Study Using FI

Considering the block direction railway object, a dangerous situation occurs when the DaKo system's block direction information is the opposite direction than the actual block direction. This situation could occur when an opposite block direction information is sent to the DaKo or when the block direction changes but the system does not transmit this information to the DaKo. Thus the critical parts of the system are the input processing parts of the LI cards and the voting part of the RPIs. These are the parts where fault-injection should be applied to assess the system's robustness.

11.4 The ProSigma FI Framework

Hardware and software failures may both occur with non-negligible probability, especially in a complex safety-critical system operating in harsh environments, and both types have a potentially huge impact on the system and on the application (railway signaling in the ProSigma case). As presented, the FI technique aims at emulating situations in which the system and its fault tolerance mechanisms face the activation of hardware and software faults, and, at the same time, collecting information on the fault activation, errors and failures caused.

The proposed FI framework has been designed to inject hardware and software faults, taking advantage of on-board scan-chain circuitry (or OCD), to emulate faults with controlled intrusiveness. The proposed framework also provides the infrastructure for collecting the experiment results automatically, allowing posterior validation of system safety requirements. In particular, the FI framework is based on the JTAG scan-chain circuitry, a *de-facto* standard implemented on a large variety of microcontrollers, including those used in safety-critical scenarios. The JTAG allows, for instance, reading values in RAM without interrupting the controller execution, and writing values in local controller registries. These are key features to both inject the faults and collect direct impact at CPU level. As an example, a bit-flip fault is injected by stopping the controller execution, reading the value of a CPU register, changing the value of a given bit (or bits), and writing back the new value to the register. The intrusiveness of such injection operation is just a few operation cycles.

11.4.1 Fault Injector Framework Architecture and Functionalities

The architecture of the FI framework, shown in Figure 11.6, is made up of several components, distributed on a host system and on the target system, namely:

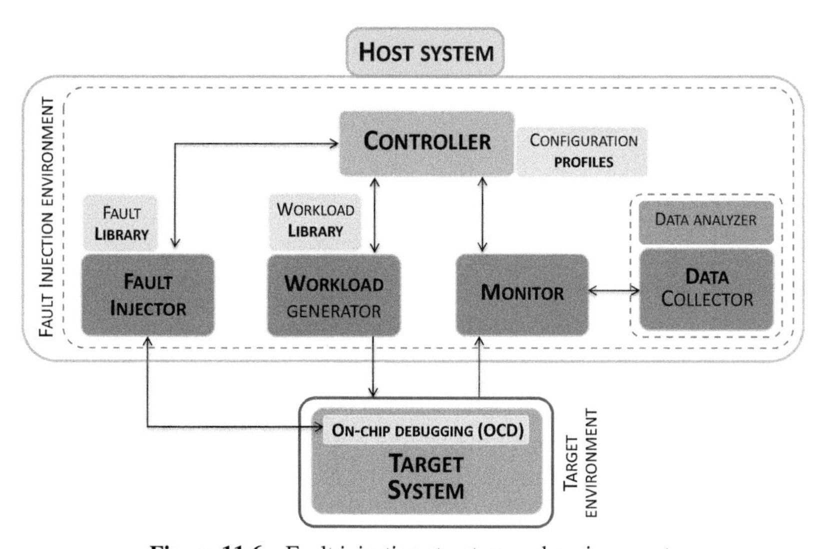

Figure 11.6 Fault injection structure and environment.

- the **fault injector component**, executing a set of instructions directly on the target system, using the OCD interface of the target system. The fault to be injected are defined in a specific module of the injector, called fault library;
- the **workload generator**, controlling the inputs to the target system. The stimuli are stored in a workload library;
- the **monitor**, which collects information about the correct functioning of the target system from the target system and its environment. The data is stored in a collection module, including a data analyzer for the user;
- the **controller module**, which orchestrates the several modules of the FI tool according to parameters specified in the form of configuration profiles.

Fault Injection campaigns consist of five phases, ranging from the definition of the faults to their injection, ending in the analysis of the results. In details:

- **Definition phase**: the user defines the faults to inject and their locations, the workload details and profiles, and the information to be monitored;
- **Set-up phase**: in this phase the user connects the FI environment installed in the host system to the target system, configure the profile of the FI campaign(s), and defines the target system requirements to be validated automatically by the system;
- **Execution phase**: in this phase the user launches one FI campaign at a time, which can be paused and resumed at any moment. A FI campaign is made of several runs, each run executing the target system (in this context the ProSigma system, or part of it) and injecting a fault (FI run, or FIR). Alternatively, runs with no fault injected are called Golden runs (GR), which are useful to observe the nominal behavior of the system;
- **Analysis phase**: this phase serves for analyzing the data collected for possible errors and failure events collected. Depending on the target system, a huge variety of analysis can be carried out;
- **Validation phase**, finally, correlating the errors and failure events, if any, to the target system requirements defined.

11.4.2 The ProSigma FI Tool (ProSigma-FIT)

The proposed framework was implemented into the ProSigma FI tool for the ProSigma system (ProSigma-FIT). A representation of the implemented FI environment is depicted in Figure 11.7. The tool can inject “bit-flips” hardware faults, i.e., emulating the flip from 0 to 1 or viceversa, in one of

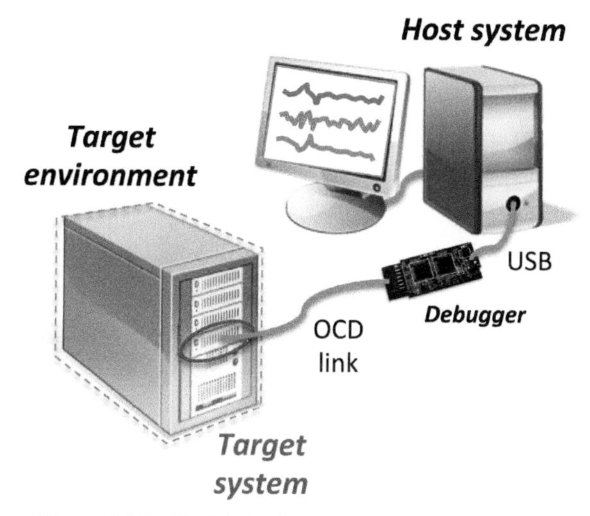

Figure 11.7 Fault injection structure and environment.

the positions of a given registers, These kinds of faults are usually caused by environmental conditions, as charged particles passing through the circuitry. ProSigma-FIT injects faults in the microcontrollers that constitute the ProSigma system, using a host system running Windows 7. The host performs the injection using a debugger communicating through USB port, an external electronic board equipped with circuitry for communicating with a given set of microcontrollers using an OCD port (JTAG in the current case).

ProSigma-FIT is developed as a Java application, and it uses an external library named OpenOCD, which eases the use of JTAG protocol by offering a set of high-level command to a user of the host system. OpenOCD is a project developed at University of Applied Sciences Augsburg [19]. The tool is made of core classes, which include objects for injecting faults and saving data into a MySQL database (*Fault Injector package*), a package for managing the FI environment and the target environment (*ProSigma Environment package*), and objects for monitoring the status of the target system and its environment (*Monitor package*).

11.5 ProSigma Safety Assessment Through FI: Experiments and Results

The ProSigma-FIT was used to assess the safety mechanisms implemented by the ProSigma system, both at hardware and software level, as a whole. The ProSigma-FIT was setup to target both CPU registers and RAM memory locations of both the targets locations (G channel of the LI card and RPI), and to performed several FI campaigns.

11.5.1 Safety Assessment of the Prosigma System: Experimental Setup

We injected hardware bit flips in the ProSigma system, namely in the G channel of the LI card (target system: TI LM3S2948 microcontroller), which is one of the TMR channels, and in the RPI, node 0 (target: TI TMS570LS3137 microcontroller), which is one of the two modules contained in a RPI card performing the voting functionality. The faults are injected using two JTAG debuggers, namely the Texas Instruments LM3S8962 and the Texas Instruments XDS100. Figure 11.8 shows a photo of the complete experimental setup.

11.5.2 Results

The ProSigma-FIT injected a total of 10,702 faults in few days. Table 11.2 presents the failure modes (system's modules level) monitored by the observers in the LI and the RPI cards. Table 11.3 shows the FI campaigns performed and presents a summary of key results. Figure 11.9 shows the distribution of the failure modes in each FI campaign.

As an example, we selected one of the ProSigma system requirements (R1) to be at validated. Due to the page limit, this chapter does not address the validation of other requirements. The requirement selected is the following:

> ***R1*** *– AFTER the INPUT status is set, the system's client must EVENTUALLY receive a message indicating the SWITCHING STATE and the CORRECT OBJECT STATE.*

Most of the faults injected in the channel G of the LI card (i.e., one of the channels of the TMR) caused effects in the ProSigma system, as shown in Table 11.3. However, as expected, the system managed to tolerate all the faults injected in the LI card. In addition to the more detailed analysis of the fault effects (especially for the ones that caused Crashes and Performance Failures) in the LI card, more comprehensive FI campaigns are needed to gain additional confidence in the system. Previous FI experiments performed for space application [3] have shown that unexpected error propagation due to shared resources such as memory may cause common mode failures.

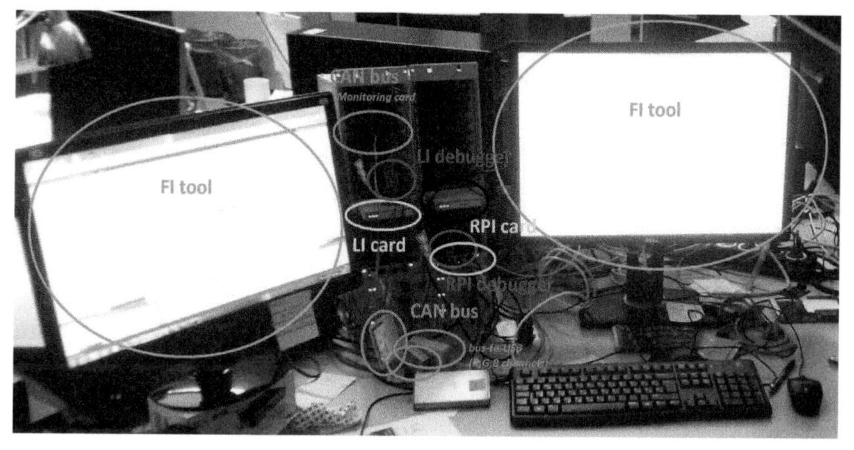

Figure 11.8 The ProSigma system and the FI tool and environment.

Table 11.2 Failure modes

Target	Observer	Failure Mode	Conditions
LI card, G channel	CAN bus	No PIT* messages (NPm)	No PIT messages from the G channel on the CAN bus for more than 3 seconds
LI card, G channel	CAN bus	No CONN** messages (NCm)	No CONN messages from the G channel on the CAN bus for more than 3 sec.
LI card, G channel	CAN bus	Performance failure (P)	PIT or CONN messages appear late on the CAN bus (latency between 1 and 3 seconds)
LI card, G channel	CAN bus	Crash (C)	No PIT and CONN messages for more than 3 sec.
RPI card, module 0	CAN bus	All the same failure modes defined for the LI card	

*PIT is a high-level protocol implemented by the ProSigma system in the LI card.
**CONN is a low-level protocol (right above the CAN messages) implemented by the ProSigma system in the LI card.

Concerning the faults injected in the RPI (voting) card, a single campaign was enough to observe Crash failures that caused the system to stop working, entering in a fail-safe state. Next FI campaigns will be focused on the comprehensive evaluation of the SW voting elements.

As shown in Figure 11.9, the faults injected caused a significant percentage of failures in the target (G channel and RPI). In particular, faults injected in the LI card caused failures in the G channel in about 30% of the

Table 11.3 Summary of FI campaign results

		Target Failures				
Campaign	# FI Runs	NCm	NPm	P	C	ProSigma Behavior
#1 (LI, registers)	674	0	15	5	152	Failure tolerated
#2 (LI, registers)	618	0	2	0	159	Failure tolerated
#3 (LI, registers)	720	0	5	1	172	Failure tolerated
#4 (LI, registers)	721	0	6	0	171	Failure tolerated
#5 (LI, RAM)	2,116	0	10	0	828	Failure tolerated
#6 (LI, RAM)	2,950	0	0	0	854	Failure tolerated
#7 (LI, RAM)	2,150	0	23	1	828	Failure tolerated
#8 (RPI, registers)	753	0	28	1	472	Safety state (Crash)
Total	10,702	0	61	7	3164	

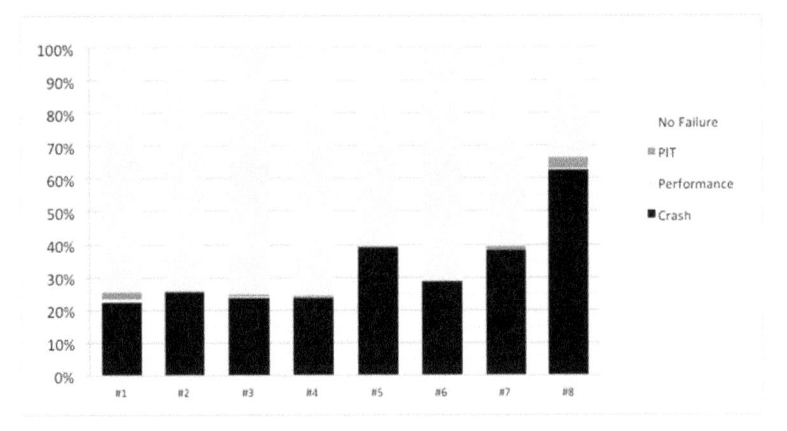

Figure 11.9 Fault injection campaign: failure modes distribution.

times, with "Crash" failures being the most frequent type, followed by "No PIT messages" and "Performance", and with "No CONN messages" failures only occurred when a Crash occurred, without any isolated occurrence. Conversely, more than 60% of the faults injected in the RPI card caused failures, most of which were "Crash"-type. We believe that such behavior is due to additional fault tolerance mechanisms contained in the TMS570LS3137 microcontroller, as the lock-step schema.

Finally, during the campaigns we measured an average injection time below 1ms (round-trip-time host-controller-host). The injection operation is hence quite invasive, being the period of the fastest microcontroller of 6.25 ns. However, the impact of the introduced latency can be tolerated by the single target system. We aim at implementing dedicated module to reduce the injection time in a future work.

11.6 Conclusion

This chapter presented a FI tool based on the JTAG technology proposed for validating a safety-critical railway signaling system, called ProSigma, a TMR system for railway trackside signaling and communication purposes. The ProSigma system has been developed at Prolan Zrt., and has been designed for being certified by the CENELEC standards as SIL 4, the most demanding level in terms of Safety availability.

The FI tool demonstrated to potentially reduce costs related to V&V activities, as it is able to highlight critical situations in which the system under test acts in a hazardous manner. The use of automated FI campaigns, focused on several components of the target system, allows to expose the system to a very large number of fault scenarios, helping gaining confidence in the safety properties of the system under validation. Results from a thorough FI campaigns are presented, illustrating the effectiveness of the FI tool and the approach in general, which confirms to be a valid instrument to help on the V&V of safety-critical system.

References

[1] Bonfiglio, V., Montecchi, L., Irrera, I., Rossi, F., Lollini, P., and Bondavalli, A. (2015). "Software Faults Emulation at Model-Level: Towards Automated Software FMEA," in *Proceedings of 2015 IEEE International Conference on Dependable Systems and Networks Workshops (DSN-W)* (New York. NY: IEEE), 133–140.

[2] European Committee for Standardization. Available at: www.cen.eu

[3] NASA. (2004). *NASA Software Safety Guidebook, NASA-GB-8719.13.*

[4] ISO. (2011). *Product development: software level. ISO 26262: Road vehicles – Functional safety 6.*

[5] Microsoft Corporation. (2014). *Resilience by design for cloud services.*

[6] Barbosa, R., Costa, D., and Madeira, H. (2006). "An empirical approach to assess software off-the-shelf components using fault injection," in *International Conference on Data Systems in Aerospace, DASIA 2006*, Berlin, Germany.

[7] Madeira, H., Some, R. (NASA), Moreira, F., Costa, D., (Critical Software), and Rennels, D. (UCLA). "Experimental evaluation of a COTS system for space applications," in *IEEE/IFIP Int. Conf. on Dependable Systems and Networks* (New York. NY: IEEE), DSN, USA, June 2002.

[8] Silva, A., Sánchez, S., Polo, O. R., and Parra, P. (2014). Injecting faults to succeed. Verification of the boot software on-board solar orbiter's energetic particle detector. *Acta Astronautica*, 95.

[9] Wei, S., Cai Bai-gen, Chen-xi, G., Jian, W., Jing-jing, W. (2010). "Research on reliability evaluation of high-speed railway train control system based on fault injection," in *Int. Conf. Environmental Science and Information Application Technology (ESIAT)*, Vol. 3, 288–293, Wuhan, China, 17–18 July.

[10] Benso, A., and Prinetto, P. (2006). *Fault Injection Techniques and Tools for Embedded Systems Reliability Evaluation.* Berlin: Springer Science & Business Media.

[11] http://www.prolan.hu/en/divisions/railway-automation/prosigma/

[12] CENELEC. (1999). *EN 50126-1:1999 Railway application, The specification and demonstration of Reliability. Availability, Maintainability and Safety (RAMS), Part 1: Basic requirements and generic process.*

[13] CENELEC. (2007). *EN 50126-2:2007 Railway applications. The specification and demonstration of Reliability, Availability, Maintainability and Safety (RAMS) – Part 2: Guide to the application of EN 50126-1 for safety.*

[14] CENELEC. (2011). *EN 50128:2011 Railway applications: Communication, signaling and processing systems – Software for railway control and protection systems.*

[15] CENELEC. (2003). *EN 50129:2003 Railway applications: Communication, signaling and processing systems – Safety related electronic systems for signaling.*

[16] Hsueh, M. C., Tsai, T. K., and Iyer, R. K. (1997). Fault injection techniques and tools. *IEEE Comput. J.* 30, 75–82.

[17] Natella, R., Cotroneo, D., and Madeira, H. (2016). "Assessing dependability with software fault injection: a survey", in *ACM Computing Surveys* (ACM: New York, NY), Vol. 48.

[18] ProSigma. *Prolan Process Control Co.* Available at: http://www.prolan.hu/en/divisions/railway-automation/prosigma/

[19] OpenODC. *University of Applied Sciences Augsburg.* Available at: http://openocd.org/documentation/

12

Robustness and Fault Injection for the Validation of Critical Systems

Nuno Laranjeiro[1], Gonçalo Pereira[1], Seyma Nur Soydemir[1], Raul Barbosa[1], Jorge Bernardino[1,2], Cristiana Areias[1,2], Nuno Antunes[1], João Carlos Cunha[1,2], Marco Vieira[1] and Henrique Madeira[1]

[1]CISUC, Department of Informatics Engineering, University of Coimbra, Portugal
[2]ISEC – Coimbra Institute of Engineering, Polytechnic Institute of Coimbra, Portugal

Critical systems are nowadays being deployed as services or web applications, and are being used to provide enterprise-level business-critical operations. These systems are supported by complex middleware, which often links different systems, and where a failure can bring in disastrous consequences for both clients and service providers. In this chapter we present a toolset that can be used to evaluate the robustness of a given system, under the following two different perspectives: i) executing robustness tests against the service's external interface (e.g., the interface with business clients) and also inner interfaces (e.g., the application-database interface); ii) emulating the presence of source code defects, on the service middleware, to understand how the presence of a defect can affect the robustness of the overall system. The toolset has been demonstrated on a set of web services, an Enterprise Resource Planning web application, and on the popular Apache HTTP server. Results show that the toolset can be easily used to disclose critical problems in web applications and to support middleware, helping developers in building and validating more reliable services.

12.1 Introduction

Web applications and services are nowadays used as the interface of many businesses to the outside world, providing services that are frequently supported by web servers and back-end databases. In these environments, a service failure can damage the complete business, potentially bringing in considerable losses for service providers. These losses might be due to lost business transactions, but can also refer to other kinds of financial losses (e.g., time to repair, human resources used to recover systems), including reputation losses [1].

The need for practical means to assess the robustness of Web-based systems (e.g., web applications or services) is supported by several studies, which show the predominance of software faults (i.e., program defects or bugs) [2–4] as the root cause of computer failures. If we consider the huge complexity of modern software, the weight of these faults will tend to increase. Web services and applications are certainly no exception, as they are normally quite complex software applications, supported by several components of also high complexity. Moreover, the current tendency of fast-paced development of software leads developers to focus on functionality, and this means that non-functional requirements, such as application robustness, are many times overlooked [5] leading to the deployment of applications holding residual bugs.

Considering the typical structure of an application built for the Web and including supporting software, we identify the following three key issues to be handled: (i) how the application behaves in the presence of **external interface faults** (e.g., invalid client inputs); (ii) how the application behaves in the presence of **inner interface faults** (e.g., invalid database data, delivered to the application); and (iii) how residual **software faults** on the supporting middleware can affect the overall system. These obviously are not exhaustive, but represent an integrative and comprehensive view that relates to the robustness of an overall web-based system and that many works tend to overlook.

Interface faults, which relate to issues in the interaction among different software components/modules [6] are quite relevant in service environments, in particular **external interface faults**, as services are highly exposed to heterogeneous or malicious clients on the Web and must be able to provide robust service to clients, even when facing invalid inputs (generated by bugs in client applications, corruptions caused by silent failures in the network, or even security attacks).

From a robustness perspective, the problem regarding interface faults is generally tackled from an external point-of-view, typically with robustness tests targeting the public interface. However, industry reports suggest that the way **applications handle incoming data at the application–database interface** is an aspect that is many times disregarded. In fact, developers many times assume that the data being handled by the application is correct, which experience shows that is not always the case. This is corroborated by industry reports [1], where the presence of poor quality data has led to severe system failures and/or huge financial losses. As mentioned, this kind of problem at the interface level is relatively well-known in the robustness testing domain, where tests using invalid inputs applied on external interfaces of many different systems have been successfully used [5, 7, 8]. However, the definition of tests at the inner interfaces (e.g., application-database) has been largely overlooked.

In general, web applications or services rely on a web server, which in practice is a software container that supports the whole application. This kind of middleware component is subject to change (as any software component nowadays built for the Web), as providers want to deploy the latest versions, where usually a number of software bugs are corrected but, at the same time, the new code brings in the potential for more bugs. Despite the popularity of this kind of component there is still little information of the behavior of **web applications in the presence of residual software faults in the web servers**. Thus, the absence of practical means to assess the behavior of web-based applications in this kind of scenario is a strong limitation for deployments where dependability is of critical importance.

In this chapter we present a toolset, summarized in the next paragraphs, that targets the three above mentioned issues and is composed of the following tools:(i) wsrbench – a tool for testing the robustness of web services; (ii) PDInjector – a tool for testing the behavior of web applications and services in presence of poor quality data; (iii) ucXception – a tool for the practical injection of software faults, demonstrated on a popular web server.

wsrbench generates and applies **external-interface** robustness tests to SOAP web services. The goal is to understand the behavior of the system being tested in the presence of invalid inputs or stressful conditions [7], possibly exposing internal errors and allowing developers to solve the identified problems. This kind of technique can be used to distinguish systems according to the number and severity of the problems disclosed, from a black-box perspective [7, 8]. wsrbench has been used, for the first time, to test the web services of a Diagnostic Centre for a Locomotive On-board Computer, with the results revealing problems that required urgent developer attention.

PDInjector was built based on an approach whose main concept revolves around the idea of **injecting poor quality data at the application–storage interface**. We inject mutated data on returning result sets from the database. The mutated data is based on typical data quality problems, which were identified in a survey of the state of the art in dirty data [9]. In short,

PDInjector replaces valid data coming from the database with poor quality data (that should be correctly handled by the application) and observe the application behavior. The tool was used to show how a major open-source Enterprise Resource Planning web application would handle invalid inputs, and was able to disclose serious bugs, not only in the code being tested, but also issues in the Object Relational Mapping (ORM) framework, and JDBC driver used by the web application.

Finally, ucXception formally describes a set of software fault injection operators, and includes a comprehensive test suite to apply these operators to emulate faults in the software and to verify its correctness. The performance of the fault injection process is optimized by compiling only the file in which a fault is injected and linking/installing that file. We illustrate the use of the tool by carrying out an experimental evaluation using the currently most popular web server, the Apache web server. The results revealed issues that require developer's attention, including potential security problems.

The structure of this chapter is as follows. Section 12.2 presents related work on robustness testing and software fault injection and Section 12.3 describes in detail our toolset for robustness testing and fault injection of services. Section 12.4 describes the operating mode of our toolset and Section 12.5 illustrates the use of the toolset in 3 case studies. Finally, Section 12.6 concludes this chapter.

12.2 Related Work

Robustness testing is a technique that allows understanding the behavior of a system when in presence of invalid input or stressful conditions [7]. The objective is to stimulate a particular system to expose possible internal errors, which will then allow developers to solve the identified problems. The technique can be used to distinguish systems according to the number and severity of the issues uncovered and has been mostly applied to external (i.e., public) interface of several systems, from a black-box perspective [7, 8]. Using robustness tests on the inner interfaces of different independent systems is something that has been largely overlooked in previous research.

Ballista [7] is a tool for robustness testing that uses a combination of acceptable and exceptional values on calls to kernel functions of operating systems. The values that are used in each call are randomly extracted from a predefined set of tests that apply to the particular data type involved in the call. The robustness of the system being tested is classified according to the CRASH scale [7], which distinguishes five failure modes. MAFALDA [8] is also a robustness testing tool that targets microkernels. In previous work we defined an approach to assess the behavior of web services in presence of mutated SOAP messages [5], which are used on web services call parameters. The services are classified according to the failures observed during the tests, using an adapted version of the CRASH scale.

Fuzzers are tools that can be used to disclose security problems in applications. In short, from a code perspective, these problems, essentially refer to the presence of code vulnerabilities (i.e., bugs in *lato senso*) or to the use of bad programming practices [10]. Although the domain is security, it is common for these tools to operate by providing erroneous data (originally random) to the applications' interfaces. Our toolset includes a robustness testing tool, wsrbench, which essentially operates by generating wrong data that is used on call parameters for web service operations.

Data quality has been defined in a huge number of different ways in the literature [11]. The ISO/IEC 25012 standard defines it as "the degree to which a set of characteristics of data fulfills requirements" [12]. A few examples of such characteristics are completeness, accuracy, or consistency [9]. The requirements mentioned in the definition express the needs and constraints that contribute to the solution of a problem [13].

Industry reports have shown the severe damage caused by the presence of poor quality data in many different contexts [14–17], with the Gartner Group identifying bad data as the main cause of failure in CRM systems [18]. The growth of the volume of data can also increase its management complexity, which can result in a higher probability of generating poor data. The current fast-changing dynamics of the Web environment can also lead to the degeneration of customer data (e.g., due to the update software components holding bugs) and this is something has actually been reported in real systems [18].

Activities such as analysis or improvement of data quality have gathered much of the attention (e.g., to perform data cleaning) of researchers and practitioners [16, 19–23]. In fact, the impact of poor data in business critical systems [24] is quite well-known. Despite this, understanding how well an application is prepared to handle the inevitable appearance of poor data has been largely disregarded. To achieve this goal, it is essential to identify representative data quality problems and to understand how to integrate them in test cases. We researched the state of the art in data quality classification and data quality problems in previous work [9], precisely to support the definition of these test cases.

The impact of erroneous data on the reliability of web services has studied in [25]. The approach includes creating an architecture view of the system under test; assessing the data quality or validity with a tool; assessing the reliability of the data and software components; defining a state machine using the architecture as basis; and calculating the system reliability. The invalid types used in the work are limited to seven issues that are already present. So the approach is limited to reliability estimation based on identified and already present issues. In previous work [26], we created a preliminary view for a testing approach using data quality problems. PDInjector uses the above reasoning to generate faults, that emulate the presence of dirty data in the database that are delivered to a running application. The complete tool uses a comprehensive set of data quality problems [9] and is based on the approach presented in [27].

Software fault injection is currently a quite mature topic, after emerging from the general area of fault injection, and gained the interest of researchers as a specific category of fault injection technique and related tools. A survey presenting a quite comprehensive perspective on software fault injection can be found in [28].

In general, the goal of injecting software faults (i.e., realistically mimicking software defects or bugs) is to assess the impact that the activation of residual bugs in specific software components (the target component) has in the rest of the system. This is useful for different purposes. For instance, software fault injection experiments can be used to provide feedback to the development process of fault-tolerant systems [29], to validate software fault tolerance mechanisms [30], to analyze how error propagates in component-based software [31], to perform dependability benchmarking of operating systems [32], to experimentally assess the risk of using legacy software components [33], or to assess recovery features in virtualized environments [8, 34].

An important question in software fault injection is how to represent (i.e., model) software faults (bugs). The concept that a software bug has unique features and is very difficult to emulate by a fault injection tool has persisted for a long time. Actually, the early works on software fault injection assume rather simple software fault models, extracted from educated guesses of developers and testers [29, 30].

A first work that tried to emulate realistic software faults based on field studies on real bugs is presented in [35]. However, the assumption in this early work on software fault injection representativeness was that real data on residual bugs found in the target system was available, in order to generate accurate fault models. Unfortunately, in general there is no data on real bugs previously found in the target systems, so the technique proposed in [35] was of limited use.

A field study on real software faults was presented in [36]. This work concluded that there is a quite short list of software fault types that represent the most frequent types of bugs found in deployed software. Actually, it was found that more than 60% of the software faults found in the field fall in just 13 fault types. This finding opened the possibility of creating a fault injection tool (G-SWIFIT) that injects the most frequent types of faults, knowing that even with a small number of fault injection operators (a fault operator injects a given fault type) it is possible to achieve a reasonably good coverage of the software faults universe.

A recent work [32] researched whether software faults injected according to the fault types identified in [36] are really representative of residual elusive software faults. The results show that in some cases a significant share (up to 72%) of injected faults cannot be considered representative of residual software faults.

Concerning the technology of fault injection tools, the tool proposed in [36] (G-SWIFIT) injects faults at the executable code level. This has the advantage of being able to inject faults in any software component, even without access to its source code. However, the precision of the fault injection operators is not ideal, as the information available at assembly level (the G-SWIFIT tool creates an assembly version from the target executable code) does not allow a perfect identification of all the code patterns where a given fault type can be injected. Despite this, G-SWIFIT has the advantage of being quite fast, as the injection process consists of changing just a few bytes in the executable code.

The tool used in [32] assumes that there is access to the source code of the target software. This allows a more precise emulation of the faults but incurs on the extra cost of requiring the compilation and linking of the target code after the injection of each fault, which can take a considerable amount of time. The ucXception tool discussed in this chapter also assumes that the source code is available but considerably optimizes the process of injecting the faults.

In the context of this chapter, it is relevant to overview some fault injection tools that can inject software faults (in fact, most of the fault injection tools found in the literature only emulate hardware faults).

JACA [37] is a source-code independent tool that has been designed to validate Java applications. It injects high-level software faults and is based on reflection to inject interface faults in Java applications at the bytecode level [38]. The goal of this tool is to use high-level programming features to corrupt attribute values, methods parameters or return values during runtime. The Java Software Fault Injection Tool [39] does not need the source code to perform the injection because it can directly mutate compiled code and it is based on the G-SWFIT [36].

The SAFE tool [32] is an application that uses Software Implemented Fault Injection (SWIFI) technique to inject realistic software faults in programs coded in C and C++. This tool uses MCPP as parser, to get the tree of code and then applies some variations to the original files (code with simple mutations) with the selected operators. SAFE implements thirteen operators, the same number as in G-SWIFT [36]. However, while SAFE implements these operators at the source code level, G-SWIFT implements them at the binary level.

The third tool described in this chapter, the ucXception software fault injection tool, and it follows an approach similar to SAFE [40] in terms of output, which consists of producing the source code files with changes made in it. However, the proposed tool was designed to optimize the maintainability and easiness of using the fault injector, which is an essential aspect to disseminate the use of software fault injection tools.

12.3 Robustness Testing and Fault Injection for the Robustness Evaluation of Services

In this section, we explain the concepts implemented by our toolset that allow the evaluation of services, based on robustness testing and fault injection. Figure 12.1 overviews our target scenario by depicting a runtime interaction between a client and a service application. The service is supported by an HTTP server and also makes use of a database. We highlight in red critical points where our tools inject faults. Faults in the client request are set by our

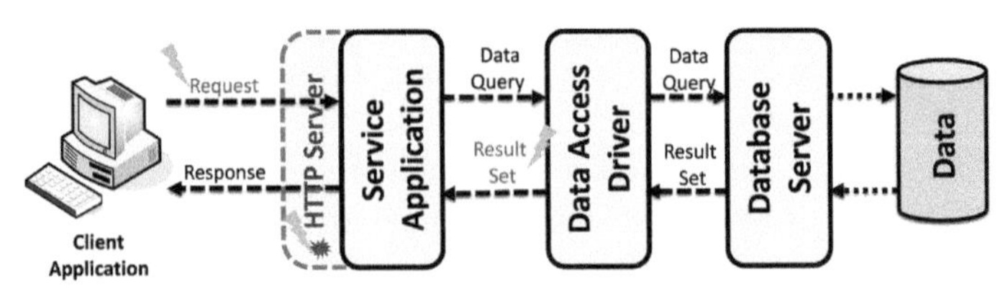

Figure 12.1 Scenario for service robustness evaluation using wsrbench, PDInjector and ucXception.

robustness testing tool for external interfaces (SOAP interfaces) – **wsrbench**; faults in result sets that are delivered to the application are injected by our robustness testing tool for inner interfaces (application-database interfaces) – **PDInjector**; and finally, we emulate the presence of software faults in the underlying middleware being used with our third tool – **ucXception**, which is capable of injecting such faults offline. As wsrbench and PDInjector are, in practice, separated by small details, we merge the explanation of both tools in the next paragraphs, explaining the differences whenever appropriate. ucXception is overviewed in the last part of this section.

12.3.1 Robustness Testing with wsrbench and PDInjector

wsrbench is a web-based tool for external interface testing of web services and is available at *wsrbench.dei.uc.pt*. In summary, it operates by sending invalid calls to web service operations. For this, the tool user must provide a WSDL document location, which describes all service operations, including input and output parameters (and data types). The tool then generates random calls based on the operation arguments data types and also domains (when provided by the user). These random calls are then sent to the service without further changes (to understand its behavior without the presence of invalid inputs). The remaining calls are mutated to include the invalid parameters that typically form the robustness tests. These mutated calls will eventually be sent to the service, so that its behavior can be assessed.

PDInjector is the tool used for testing the internal interfaces. It is based on the presence of an instrumented data access driver (e.g., a JDBC driver) that should be placed between the application, generically designated by *service application*, and the data storage system. This instrumented driver has the exact same interface of a regular data access driver, which essentially means that to use the tool, no changes to the service code, database management system, or database are required. The main idea is that the driver intercepts all calls to the database management system and, at specific moments, simulates the presence of poor quality data, by replacing the original data coming from the database with poor quality data. The goal is to understand if the application can handle the mutated data coming from the database in a robust manner or if, otherwise the service is poorly built and cannot tolerate the presence of dirty data (e.g., by becoming unavailable or throwing unexpected exceptions when processing the data).

Both tools wsrbench and PDInjector are prepared to operate according to the following sequential phases:

1. **Warm-up**: Valid client requests are issued to the service application and the goal is just to warm-up the system to reach typical operational conditions;
2. **Injection**: Invalid inputs are injected before the application code is reached(invalid inputs are generated at the client by wsrbench, and at the driver by PDInjector);
3. **Analysis**: The service behavior is analyzed by examining the responses produced by the system.

During the first two phases, PDInjector assumes there is a workload generation client that is able to place valid requests on the system, as requests must be generated by some outside entity. Since wsrbench is placed at the client-side, it is able to generate those valid requests. These requests are then used as basis to perform different functions according to the phase being executed.

During the **warm-up phase** all calls pass from the client and reach the server and there is no injection of invalid data by wsrbench. Similarly, PDInjector intercepts all data access calls, but does not inject any mutated data during this phase. As mentioned, the goal is to let the system warm-up and reach typical working conditions.

During the **injection phase**, we replace genuine data generated by the client in wsrbench, or coming from the database in PDInjector, with data that, for the particular data type and value being handled, represents a robustness problem. The types of problems that should be emulated by our tools were based on previous studies on robustness testing for wsrbench and also on a survey in data quality classification, where we identified representative data quality problems (e.g., misspellings, abbreviations, empty data, extraneous data) associated with common data types (e.g., text, numbers) [9]. The total number of mutations is quite large and, as such, we present a subset in Table 12.1 (detailed versions can be consulted in [5] and [41]).

The injection of invalid/mutated data can be performed once per each client call, since the goal is to understand how faulty data can affect the execution of that particular operation. Despite this, it is also possible to inject a given number of faults during the execution of a service operation (which we have followed in our experiments with PDInjector). Although injecting several faults may lead to difficulties in understanding the exact causes of a failure, it is frequently the typical choice in the robustness testing domain due to its simplicity and ability to disclose problems.

It is desirable that all public operations should be tested, but this obviously depends on the test being executed. For instance, a developer may be only

Table 12.1 Examples of Robustness and poor data quality mutations

Data Type	Robustness Mutations	Data Quality Mutations
String	Replace by null	Replace by null
	Replace by empty	Replace by empty
	Replace by string with nonprintable characters	Replace a word by a misspelled word (Dictionary-based) or, if no match, use a random single edit operation (insertion; deletion; substitution of a single character; or transposition of two adjacent characters) over a randomly selected word
	Add nonprintable characters to the string	Add whitespace in a leading or trailing position, or between words (random choice)
	Replace by alphanumeric string	Add extraneous data in leading, trailing, or random position (random choice)
Integer	Replace by MAX_INTEGER	Add one numeric character
	Replace by MAX_INTEGER – 1	Set to zero
	Replace by MAX_DOMAIN + 1	Remove one random numeric character
	Subtract 1	Flip sign
...		

interested in testing a few operations). In any case, for each operation being tested, each of the operation parameters (wsrbench) or of the data access points (PDInjector) present in the code should also be tested in this phase. This desirable execution profile of the injection phase is represented in Figure 12.2.

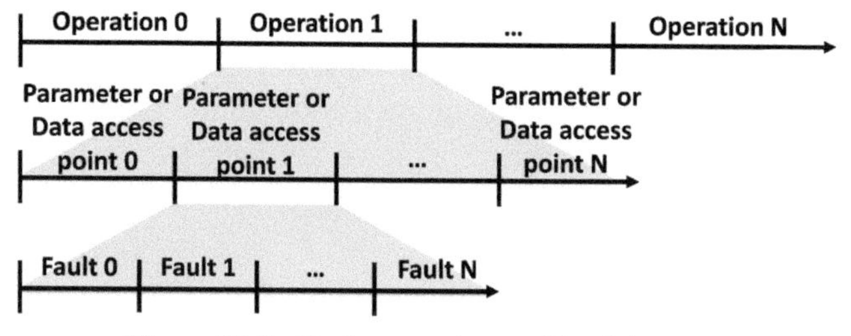

Figure 12.2 Basic execution profile of the tests.

Covering all data access points naturally depends on the client workload, which must provide adequate coverage. This aspect is currently out of the scope of our toolset. In some cases, data access points may be shared by different service operations. Even in such cases, it is desirable to invoke the different service operations, as it will exercise different areas of the code, thus having the ability to disclose different problems. Each data access point (or operation parameter, in the case of wsrbench) should be tested by PDInjector with all predefined poor data faults.

The injection phase could be automatically configured to stop when a given number of data access points (or operations/parameters) has been covered by the tests. After starting a test, wsrbench stops when all operations/parameters have been tested. In the case of PDInjector, we currently determine that a test should stop when either the client action has concluded (i.e., a response is delivered to the client) or when a failure is detected.

The last phase refers to the **analysis** of the tests results, which includes classifying any observed failures (e.g., using a failure mode scale, such as CRASH [7]) and also understanding the location and origin of the problems disclosed during the tests. Performing this analysis step requires source code access to understand if the output of the tests is a bug, what is the exact location (note that an value delivered to a specific point in the code may only be improperly used in another part of the code) and why it is a problem (so that it is corrected). If the developers want to classify the service behavior it is possible to use the CRASH scale, which classifies the severity of failures in Catastrophic, Restart, Abort, Silent, or Hindering. This kind of classification is only necessary if the developer needs to, for instance, prioritize bug fixing.

12.3.2 Emulating Software Faults with ucXception

This section describes the operating mode of the ucXception software fault injection tool, which aims at simplifying and generalizing the process of emulating software faults. ucXception modifies the source code of programs by applying software fault emulation operators over abstract syntax tree and produces software patches in an automated way. The existing specifications of emulation operators in the literature, target the emulation of software faults in binary code, and source-level operators have not been properly specified. ucXception currently uses a refined and adapted version of the existing specifications to work at the source-level.

This fault injector tool injects faults in code and the emulated faults resemble real software faults made by actual developers, which might lead to bugs. This tool was created in Java, using Eclipse CDT Plugin, and is able to inject faults in C code by following the Software Implemented Fault Injection (SWIFI) technique.

Table 12.2 shows the most important fault operators of ucXception. To apply each of these operators, there is a set of rules, which we name constraints, that assure the realism of the injected faults. These constraints were created based on the observation of real bugs [10]. The injection is location-based, because only the code locations that validate all constraints related to each of the operator, can be used to inject faults realistically. Table 12.3 shows all the constraints that are used with the operators. Each of the operators can use one or more constraints and an important aspect is that all related constraints of a given operator must be valid so that it can be applied in some location of the code.

The main tasks performed by the ucXception software fault injection tool are as follows:

1. Read source code;
2. Create the Abstract Syntax Tree (AST);
3. Verify all constraints related to the current operator;
4. Apply the operator in the AST (if the validity of the constraints is verified);
5. Create the patch with the modifications, comparing modified code with the initially source code.

Table 12.2 Fault emulation operators

Operators	Description
MFC	Missing function call
MIA	Missing if construct around statements
MIEB	Missing if construct plus statements plus else before statements
MIFS	Missing if construct and surrounded statements
MLAC	Missing and sub-expr. in logical expression used in branch condition
MLOC	Missing or sub-expr. in logical expression used in branch condition
MLPA	Missing localized part of the algorithm
MVAE	Missing variable assignment with an expression
MVAV	Missing variable assignment with a value
MVIV	Missing variable initialization with a value
WAEP	Wrong arithmetic expression in parameters of function call
WPFV	Wrong variable used in parameter of function call
WVAV	Wrong value assigned to a variable

Table 12.3 Fault emulation constraints

Constraints	Description
C01	Return value of the function must not be used
C02	Call/Assignment/The if construct/The statements must not be the only statement in the block
C03	Variable must be inside stack frame
C04	Must be the first assignment for that variable in the module
C05	Assignment must not be inside a loop
C06	Assignment must not be part of a for construct
C07	Must not be the first assignment for that variable in the module
C08	The if construct must not be associated to an else construct
C09	Statements must not include more than five statements and not include loops
C10	Statements are in the same block, do not include more than five statements, nor loops
C11	There must be at least two variables in this module
C12	Must have at least two branch conditions
C13	The if construct must be associated to an else construct

The fault injector beings by reading the source code, which consists of files typically coded in C or C++. The code is then analyzed by the Eclipse CDT plugin and an AST tree is built. In order to inject a fault, the injector searches for the node where it can be injected (it evaluates the truthfulness of all constraints of the particular operator), and modifies it, according to operator specification. After that, the AST is rewritten, by reading the modified code. The output of the ucXception tool is a set of individual patches, with each patch corresponding to the emulation of a particular bug. Thus, each patch can be applied in an experimental evaluation to emulate the presence of a bug at a particular code location.

12.4 Case Studies

In this section we describe three case studies that illustrate the application of our toolset. The first case study illustrates the usefulness of wsrbench in testing the public interface of web services that serve to diagnose a Locomotive On-board computer; the second case study focuses on inner interface testing of a web based Enterprise Resource Planning system with the use of PDInjector; and finally, the third case study illustrates the ucXception software fault injection tool by targeting popular middleware for web-based systems – the Apache HTTP server.

12.4.1 External Interface Testing: Case Study #1

MFB is a Locomotive On-board Computer (LOC) used for data acquisition. The on-board computer continuously provides data to the dispatching center regarding current position, operation status and mechanical parameters of the traction vehicle. The functions of the device are the following.

- Automatically connect to the main supervisory system of traction vehicles and engine drivers.
- Management of the electronic logbook.
- Supervise of the electric traction energy on diesel and electric locomotives.
- Automatically measure the fuel flowing to and from the feeder, sending reports to the dispatch center in case of unaccounted fuel consumption.
- Supervise the engine's activity, support the engine's diagnostic and maintenance, collect digital signals supplied by the relay contactors, and count the operating hours of the mechanical machinery.

The MFB Diagnostic Centre is composed of a set of SOAP web services, implemented in PHP, that run on an Debian Linux server and use a PostgreSQL database. We used wsrbench to perform robustness tests on the web services of the MFB Diagnostic Centre. The robustness tests of the tool were complemented with manual code inspection, which focused on the analysis of the code in terms of robustness and also security.

Results for case study #1

In total, 6 services and 53 inputs were tested with the wsrbench tool, which resulted in a total of 473 tests performed. The following items highlight the main findings:

- 37 tests disclosed robustness problems.
- 7 tests results were inconclusive and were checked manually.
- 15 out of the 47 inconclusive test cases revealed the presence of real problems, requiring developer attention.
- The final results showed robustness problems in $\sim$38% of the inputs (20 out of 53).

Not all the revealed issues led directly to hang or crash failures, but these issues show that inputs are incorrectly handled, and therefore represent bug prone code. These issues also mean that there was improper preparation of the code to deal with unexpected inputs. Such cases may lead to severe problems under special operational conditions, and also hinder the maintainability of the system.

12.4.2 Inner Interface Testing: Case Study #2

In this case study, we opted to test a business-critical pure web-based application for Enterprise Resource Planning [27]. We selected a well-known widely used commercial open source ERP business solution for enterprises. This ERP is a world leader in its category, with, at the time of writing, about 2.5 million downloads. It allows managing a whole business, and supports typical business processes such as sales, manufacturing, or finance. As we are not allowed to disclose the name of the tool, we name it ERPx. ERPx requires a database, which we opted to be PostgreSQL 9.3, and a web server for which we chose the popular Apache Tomcat 7.0.68. As we had the intention to repeat the tests, besides using a web browser, we recorded and later replayed user operations on the browser (when interacting with ERPx) with SikuliX 1.1.0.

ERPx is a very large application and, as such, we opted to select a few test cases, which should suffice to illustrate the usefulness of PDInjector. We considered the CRUD model [42] for selecting operations with different profiles: CREATE, READ, UPDATE and DELETE. An important aspect is that all the selected test cases are quite complex and are mostly composed of read operations, but we classified them according to their main goal. Our intention was to obtain a good mix between operations that potentially have distinct data access patterns or are built in differently. Table 12.4 presents the operations that we selected for testing, how they map to the CRUD model, and a letter (A, B, C, D, and E) that identifies a type of failure uncovered during testing. The uncovered failures are discussed in the next section.

Results for case study #2

As we can see in Table 12.4, we were able to disclose failures in all operations tested. The uncovered issues were discovered at the following three distinct parts of the system: (i) the application code; (ii) in the widely used Object-Relational Mapping framework used by ERPx; and (iii) in the popular PostgreSQL driver code.

Table 12.4 Overview of the tests and results for case study #2

Operation Name	Type (CRUD)	Failure Reference (See Table V)
Login	R	A, B, C, D
Create Organization	C	A, C, D
Create a new User	C	A, B, C, D
Create a new Role	C	A, B, C, D
Create Product	C	A, B, C, D, E
Delete Product	D	A, B, C, D
Update Product	U	A, B, C, D
Export Product Categories	R	A, B, D

Table 12.5 presents an excerpt of the issues uncovered during the tests, which were selected due to their manifestation in different forms and due to their location in different structural parts of the system (as visible in Table 12.4). All of these examples are problematic, even in those cases where no message was shown to the user, as eventually the application became unusable.

Failure A mostly occurred whenever the data was mutated to null. However, in the case of the example, it is a mutated variable value (variable *referenceID*) that causes an access to the database to return null. This null value is then used without being checked, triggering a *NullPointerException* (the developer could first check the value to avoid the exception). This internal exception triggers a *TemplateModelException* that becomes visible to the user in an alert box.

Table 12.5 Selected cases from case study #2

Ref	Root Exception Triggered	Location	Last Mutation	External Behavior
A	NullPointerException	Application	changeToOppositeCase	TemplateModelException reported to the user
B	ClassNotFoundException	Application	add Extraneous	No message displayed to the user
C	PSQLException	Application	replaceBySQLString	Application error message disclosing table row contents
D	StringIndexOutOfBounds	JPA Middleware	replaceByEmptyString	Application error message stating String index out of range
E	ArrayIndexOutOfBounds	JDBC Driver	addCharactersToString	No message displayed to the user

ERPx loads a few classes dynamically, and *Failure B* occurs when one of the class names is incorrect (due to the injection of a mutation). We do not have enough information to state if this is the right design choice (i.e., having dynamic loading of classes), however disclosing this issue can actually help programmers understanding if this is a good design decision and especially how the application is prepared to handle this kind of situation. Anyway, the user should be informed in the event of an error (especially if it renders the application unusable), which did not happen during our tests. This suggests that the application's error handling mechanisms have space for improvement.

Failure C is a critical example. It actually is a second order SQL Injection problem, where malicious data present in the database is improperly used to build an SQL query. A malicious user might be able to obtain sensitive information, as the information coming from the database is not sanitized by the service. This shows the potential of PDInjector to disclose security problems. In addition to this, and although the error messaging system of the service was correctly triggered, the actual message displayed to the user discloses the contents of an entire database table row, which should not happen.

Failure D is an interesting case, where the tool disclosed fragility in the implementation of the Object-Relational Mapping framework used by ERPx. In this particular case, the ORM framework accesses the first character of a string and fails as the string had turned empty due to the mutation applied. It is interesting because the framework previously checks if the string is null, but it does not check if it is empty and then immediately accesses the first character. This triggers a *StringIndexOutOfBoundsException*. This is an implementation flaw, quite similar to the one described in the next paragraph (which already received a correction from the developer community).

Failure E is triggered when adding characters that include a single quote to a string. This is a bug in the driver being used in the experiments (PostgreSQL JDBC Driver 9.4–1201) that has been reported [43] and fixed in version 9.4–1204. In summary, the code fails to find the closing single quote and returns the position of the last character in the query as the end of the string. The issue here is that in another part of the driver, the code does not expect this behavior, and the outcome is an access to a position that is one place after the end.

Before executing the tests we were expecting to find a few application-level issues, but PDInjector was able to find issues at the middleware level (in fact at two middleware levels – ORM framework and JDBC driver). The fact that the middleware is widely used and also tested highlights the usefulness of the tool in disclosing issues in applications experiencing unexpected conditions. Moreover, the potential to find problems that go beyond aborted executions with exceptions (or wrong messages presented to the user) and that can actually represent security problems, further emphasizes value of this type of testing for application architects and developers.

12.4.3 Injecting Software Faults in Service Middleware: Case Study #3

In this case study, we illustrate the application of the ucXception software fault injection tool to the Apache Web server [44]. The main idea is to have the Apache Web server installed, allowing the use of HTML and PHP pages. For this, along with specific typical configurations, we installed the Apache2 HTTP Server (version 2.4.12); the Apache Portable Runtime; Perl Compatible Regular Expressions (PCRE); and PHP: Hypertext Preprocessor (PHP5, including the package libapache2-mod-php5).

We are also using the APache eXtenSion tool (APXS), so that we can install new modules (e.g., patched modules) without the need to recompile the complete Apache server. As a result, APXS makes it possible to execute a large number of experiments in a reasonable amount of time. The experiments were carried out in a specific environment, with faults being injected mod_rewrite. The goal is to demonstrate that the fault injector tool really works and that the injected faults produce effects on the Apache Web Server.

To verify the integrity of the target environment after the injection of each fault we perform the following tests:

- **Apache:** This is the most basic test that should be done at Apache after installation and consists of sending an HTTP request to *index.html.* The response is normally a page with information about the Apache installation, with the string: “It works”;
- **PHPInfo:** This test is carried out to check the operation of PHP. It provides a large amount of information to the user, including configuration settings, PHP version, OS version;
- **Out:** Request holding parameters that will be shown at the response;
- **PHPBench:** Verify and measure the time that certain PHP functions take related with the manipulation of strings and arrays;.

12.4.4 Results for Case Study #3

In order to inject faults, our injection tool analyses the source code nodes to identify where each fault operator can be applied (as mentioned, all constraints associated with a particular fault emulation operator need to be valid at a particular location). After evaluating the code of the Apache module "mod_rewrite", the tool identified 1474 locations in this component where faults could be injected. As previously mentioned, ucXception produces one patch file each time a software fault emulation operator is injected. Thus, the output of the tool is a set of patch files, each of which is applied to the original source code file in order to generate a version with the target program with the software fault. Table 12.6 shows the number of patch files that the tool produced for mod_rewrite.

It is important to refer that the number of patches created by the application of the MLPA operator is quite large, as it removes a small part of the algorithm, consisting of any combination of up to five function calls and/or statements. As a result, the application of this operator produced over one third of the patch files. Also, it is worth mentioning that only one patch was created through the application of operator WPFV. This operator consists of the replacement of a function parameter by another that must be of the same type. This restriction is related to the environment in which the fault is injected, since its lack could trigger compilation errors. Thus, the existence of only one patch is related to the need of the changed parameter to be exactly of the same type of the initial parameter. In C and C++ languages, developers typically create their own structures of a specific type, to be easier to represent the data in the application. Due to this, the types of variables used are quite different in a function and their number of occurrence is low.

The injected faults produced anomalous effects in 213 trials, corresponding to 14.45% of injected faults. In our context, a fault has no effect whenever the output is equal to the one in a fault-free program. The fault has effect if there is no output, if the output is incorrect, or if the output is corrupted. In our experiments, the output refers to the HTTP responses received from the Apache web server. The number of experiments is a result of the constraints that need to be evaluated at each potential location in the

Table 12.6 Number of patches for mod_rewrite

Operator	MIFS	MLAC	MFC	MIA	MLOC	MLPA	MVAE	MVAV	MIEB	MVIV	WVAV	WAEP	WPFV	Total
Nr. of Patches	239	79	162	260	41	526	25	15	54	14	24	34	1	1474

code. If all the constraints of one operator are valid, then the operator will be applied.

The behaviors observed during the different tests were evaluated in three steps. In the first step we selected only unique behaviors, in the second step we analyzed and compared the selected behaviors to create a classification scheme, and in the final step, a script evaluated all the behaviors, one by one, through the classification. The behaviors observed were grouped in the following three categories (failure modes):

- **Correct**: The Web server shows a correct behavior;
- **Wrong output**: The Web server returns incorrect information, or no information at all;
- **Apache error**: Refers to the errors directly related with the behavior of apache2, such as bad request, not found with/without correct url, Internal Server Error, Ok, and Forbidden. After classifying each experiment, we further divided the initial set of Apache errors into 9 distinct categories, in order to understand, in a more comprehensive manner, the diverse failure modes of the Apache web server in the presence of software faults. Table 12.7 shows each of the nine identified categories and their respective numeric reference.

Table 12.8 summarizes the results obtained during our experimental campaign, highlighting and counting the different anomalous behaviors found during the experiments. C, W, and A in the table, respectively correspond to Correct, Wrong output, and Apache error.

We can see that even the basic Apache operation is affected by the introduced faults. However, there is no rule for rewriting of files when they are available in the directory, which is the case of Apache "It works!".

Table 12.7 Types of observed behaviors

#	Description
1	Bad request
2	Empty
3	Forbidden
4	Found
5	Internal Server Error
6	Not found – url OK
7	Not found – wrong url
8	Apache error – Ok
9	Wrong output

Table 12.8 Results by behavior

#	Apache	PHPInfo	Out	PHPBench	Number
1	W	W	W	W	86
2	C	C	A	C	61
3	A	A	A	A	25
4	C	C	W	C	19
5	W	W	C	W	19
6	C	W	A	W	1
7	A	W	W	W	1
8	A	A	A	C	1

A, Apache Error; C, Correct; W, Wrong Output.

Figure 12.3 shows the number of anomalous effects produced by patch type, and according to the type of test executed.

It is quite visible in Figure 12.3 that the operator that causes the most failures is MLPA, but this is a result of the relatively higher number of patch files created, as explained earlier. Figure 12.4 shows a detailed view of the number of different effects observed during the tests, with respect to the different types of tests executed.

As we can see in Figure 12.4, the most frequent behavior is an *Empty* response, which occurs when the Apache web server responds with an empty HTML body. The remaining most frequent behaviors are *Internal Server Error* and *Wrong output*. There are 2 anomalous behaviors that occur with a quite low frequency, in particular *Forbidden* and a *Not found - wrong url*

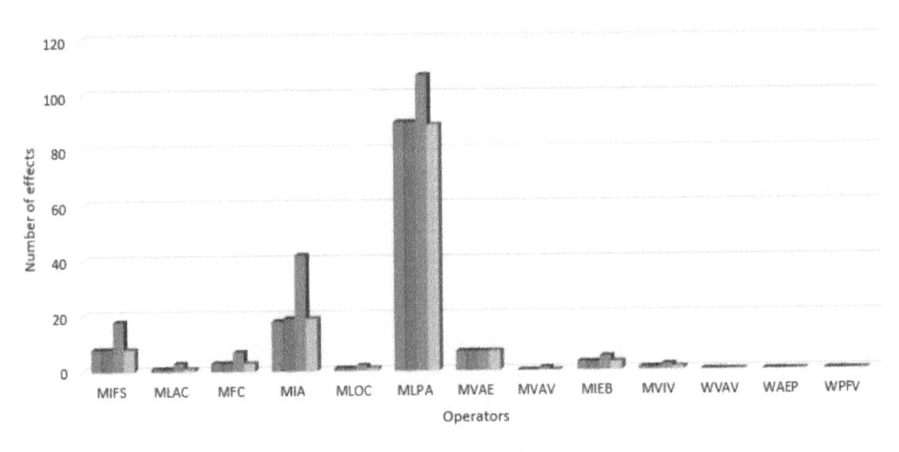

Figure 12.3 Anomalous effects by type of patch.

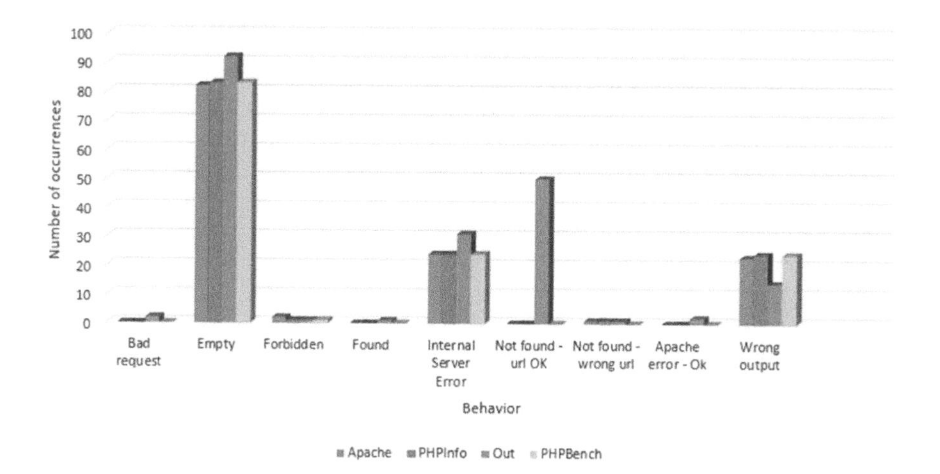

Figure 12.4 Effects by behavior.

(having an absolute frequency of two and one, respectively). Internal server errors are sent by Web servers to signal errors occurring at the server-side, while a wrong output consists of an incorrect result being sent to the user in the HTML body.

As a side note, the results obtained in PHPInfo and PHPBench tests were similar. This occurs because the PHPInfo and the PHPBench tests are based in two pages with different PHP code, but the rewriting process for both is the same. Thus, these two tests involve the same principles, the same workflow in Apache. Regarding the results of tests with some parameters, we observed that four anomalous behaviors just occur in this test, *Not found - url ok*, *Bad Request*, *OK* and *Found*. As might be expected, these are the tests where there are more errors due to a greater use of module "mod_rewrite". All the requests are rewritten, and the parameters are presented on a page in PHP.

The highest number of anomalous behaviors was observed during the *Out* test. Moreover, we should note that the error *Not found - url ok* occurs often. This behavior means that the request is done correctly, and received properly by the Apache server, but actually it is not rewritten correctly, and because there is no file with that name and in that directory, returns *Not Found*.

As a final note, in some cases, instead of obtaining the right result, it is possible to obtain the exact path where the HTML or PHP files are located, the *DocumentRoot* of Apache. This can be a serious security problem and it occurred with, for example, one application of operator *MIEB* and one of *MIFS*. There are also two cases, applying the operator *MIA*, where it is possible to see a page with PHP code, instead of the result of the execution of code. More than suggesting that a problem has occurred during the loading of the PHP module, it may represent a potential security problem, as a user will have access to internal code details. This also shows that a single fault introduced into the Apache module can affect the operation of other, as is the case of the PHP module.

12.5 Conclusion

In this chapter, we presented a toolset, composed of three tools: *wsrbench* and *PDInjector*, for robustness testing of external and inner interfaces, respectively; and *ucXception* for the injection of software faults. The tools were described from a functional perspective and their use was illustrated in three case studies. In particular, *wsrbench* was used to assess the robustness of a set of web services used in a safety-critical environment, *PDInjector* was used to assess the robustness of a web application used in business-critical environments, and finally *ucXception* was used to inject software faults in the very popular web server Apache HTTP server. The tests disclosed several different failures, including bugs at the application-level and supporting middleware for database access (i.e., JPA implementation and the JDBC driver used) and also in the web server tested. Some of the issues found were related with security problems. In future work we intend to further integrate the tools and research ways of automating the tests, possibly resorting to machine learning algorithms to analyze the behavior of the systems being tested.

References

[1] Loshin, D. (2011). Evaluating Business Impacts of Poor Data Quality. *Inform. Quality J.*

[2] Lee, I., and Iyer, R. K. (1995). Software Dependability in the Tandem GUARDIAN System. *IEEE Trans. Softw. Eng.* 21, 455–467.

[3] Kalyanakrishnam, M., Kalbarczyk, Z., and Iyer, R. (1999). "Failure Data Analysis of a LAN of Windows NT Based Computers," in *Proceedings of the 18th IEEE Symposium on Reliable Distributed Systems*, 178, IEEE Computer Society.

[4] Rodríguez, M., Albinet, A., and Arlat, J. (2002). "MAFALDA-RT: A Tool for Dependability Assessment of Real-Time Systems," in *The 2002*

International Conference on Dependable Systems and Networks (DSN 2002), 267–272, IEEE Computer Society.

[5] Laranjeiro, N., Vieira, M., and Madeira, H. (2012). A Robustness Testing Approach for SOAP Web Services. *JISA* 3, 215–232.

[6] Weyuker, E.J. (1998). Testing component-based software: a cautionary tale. *Softw. IEEE* 15, 54–59.

[7] Koopman, P., and DeVale, J. (1999). "Comparing the robustness of POSIX operating systems," in *Twenty-Ninth Annual International Symposium on Fault-Tolerant Computing*, 30–37.

[8] Rodríguez, M., Salles, F., Fabre, J.-C., and Arlat, J. (1999). "MAFALDA: Microkernel Assessment by Fault Injection and Design Aid," in *The Third European Dependable Computing Conference on Dependable Computing* (Berlin: Springer-Verlag), pp. 143–160.

[9] Laranjeiro, N., Nur Soydemir, S., and Bernardino, J. (2015). "A Survey on Data Quality: Classifying Poor Data," in *The 21st IEEE Pacific Rim International Symposium on Dependable Computing (PRDC 2015)*, IEEE Computer Society, Zhangjiajie, China.

[10] Antunes, J. and Neves, N. (2012). "Recycling Test Cases to Detect Security Vulnerabilities," in *IEEE 23rd International Symposium on Software Reliability Engineering (ISSRE 2012)*, pp. 231–240, IEEE Computer Society, Washington, DC, USA.

[11] Batini, C., Palmonari, M., Viscusi, G. (2014). Opening thande Closed World: A Survey of Information Quality Research in the Wild, in *The Philosophy of Information Quality* (Berlin: Springer International Publishing), 43–73.

[12] ISO/IEC (2008). *Software engineering – Software product Quality Requirements and Evaluation (SQuaRE) – Data quality model. ISO/IEC.*

[13] *SWEBOK V3 Guide IEEE Computer Society.*

[14] Ge, M. and Helfert, M. (2007). "A Review of Information Quality Research – Develop a Research Agenda," in *12th International Conference on Information Quality*, pp. 76–91, Cambridge, MA, USA.

[15] Pipino, L. L., Lee, Y. W., and Wang, R. Y. (2002). Data quality assessment. *Commun. ACM* 45, 211–218.

[16] Loshin, D. (2010). *The practitioner's guide to data quality improvement.* Burlington, MA: Morgan Kaufmann.

[17] Quality, E. D. (2015). *The data quality benchmark report.* Experian Data Quality.

[18] Marsh, R. (2005). Drowning in dirty data It's time to sink or swim: A four-stage methodology for total data quality management. *J. Database Market. Cust. Strat. Manage*. 12, 105–112.

[19] Caro, A., Calero, C., Mendes, E., and Piattini, M. (2007). A Probabilistic Approach to Web Portal's Data Quality Evaluation, in *6th International Conference on the Quality of Information and Communications Technology, 2007, QUATIC 2007* (New York, NY: IEEE). 143–153.

[20] Xiaojuan, B., Shurong, N., Zhaolin, X., and Peng, C. (2008). Novel method for the evaluation of data quality based on fuzzy control. *J. Syst. Eng. Electron*. 19, 606–610.

[21] Bergdahl, M., Ehling, M., Elvers, E., Földesi, E., Körner, T., Kron, A., Lohauß, P., Mag, K., Morais, V., and Nimmergut, A. (2007). *Handbook on Data Quality Assessment Methods and Tools*, Wiesbaden.

[22] Choi, O.-H., Lim, J.-E., Na, H.-S., Seong, K.-J., and Baik, D.-K. (2008). "An Efficient Method of Data Quality Evaluation Using Metadata Registry," in *Advanced Software Engineering and Its Applications, 2008, ASEA 2008* (New York, NY: IEEE), 9–12.

[23] Galhardas, H., Florescu, D., and Shasha, D. (2001). "Declarative Data Cleaning: Language, Model, and Algorithms," in In VLDB (New York, NY: IEEE), pp. 371–380.

[24] Haug, A., Zachariassen, F., Liempd, D. van (2011). The costs of poor data quality. *J. Ind. Eng. Manage*. 4.

[25] Musial, E. and Chen, M.-H. (2012). "Effect of Data Validity on the Reliability of Data-centric Web Service," in *IEEE 19th International Conference onWeb Services (ICWS), 2012*, Honolulu, HI, USA.

[26] Ivaki, N., Laranjeiro, N., and Vieira, M. (2013). "Towards Evaluating the Impact of Data Quality on Service Applications," in *Workshop on Reliability and Security Data Analysis (RSDA 2013) co-located with The 43rd Annual IEEE/IFIP International Conference on Dependable Systems and Networks (DSN 2013)*, IEEE Computer Society, Budapest, Hungary.

[27] Laranjeiro, N., Soydemir, S. N., and Bernardino, J. (2016). Testing Web Applications Using Poor Quality Data," in *Latin-American Symposium on Dependable Computing (LADC 2016)*. IEEE Computer Society, Cali, Colombia.

[28] Natella, R., Cotroneo, D., and Madeira, H. S. (2016). Assessing Dependability with Software Fault Injection: a Survey. *ACM Comput. Surv*. 48, 44:1–44:55.

[29] Ng, W.T., and Chen, P.M. (2001). The Design and Verification of the Rio File Cache. *IEEE Trans. Comput*. 50, 322–337.

[30] Voas, E., Charron, F., McGraw, G., Miller, K., and Friedman, M. (1997). Predicting how badly "good" software can behave. *IEEE Softw.* 14, 73–83.

[31] Hiller, M., Jhumka, A., Suri, N. (2001). "An approach for analysing the propagation of data errors in software," in *2001 International Conference on Dependable Systems and Networks* (New York, NY: IEEE), 161–170.

[32] Natella, R., Cotroneo, D., Duraes, J. A., and Madeira, H.S. (2013). On Fault Representativeness of Software Fault Injection. *IEEE Trans. Softw. Eng.* 39, 80–96.

[33] Moraes, R., Duraes, J., Barbosa, R., Martins, E., and Madeira, H. (2007). "Experimental risk assessment and comparison using software fault injection," in *37th Annual IEEE/IFIP International Conference on Dependable Systems and Networks (DSN'07)* (New York, NY: IEEE), 512–521.

[34] Cerveira, F., Barbosa, R., Madeira, H., and Araujo, F. (2015). "Recovery for Virtualized Environments," in *2015 11th European Dependable Computing Conference (EDCC)* (New York, NY: IEEE), pp. 25–36.

[35] Christmansson, J., and Chillarege, R. (1996). "Generation of an error set that emulates software faults based on field data," in *Proceedings of Annual Symposium on Fault Tolerant Computing* (New York, NY: IEEE), 304–313. IEEE.

[36] Durães, J. and Madeira, H. (2006). Emulation of software faults: a field data study and a practical approach. *IEEE Trans. Softw. Eng.* 32, 849–867.

[37] Moraes, R. L. de O., and Martins, E. (2003). "Jaca – a software fault injection tool," in *Proceedings 2003 International Conference on Dependable Systems and Networks* (New York, NY: IEEE), 667–667.

[38] Martins, E., Rubira, C. M. F., and Leme, N. G. M. (2002). "Jaca: a reflective fault injection tool based on patterns," in *Proceedings International Conference on Dependable Systems and Networks* (New York, NY: IEEE), pp. 483–487.

[39] Sanches, B. P., Basso, T., and Moraes, R. (2011). "J-SWFIT: A Java Software Fault Injection Tool," in *2011 5th Latin-American Symposium on Dependable Computing* (New York, NY: ACM Library), pp. 106–115.

[40] Natella, R. (2011). *Achieving Representative Faultloads in Software Fault Injection.*

[41] Laranjeiro, N., Seyma, N. S., and Jorge, B. (2016) *Poor Data Injector Toolset*. Available at: http://eden.dei.uc.pt/~cnl/papers/2016-ladc.zip

[42] Martin, J. (1983). *Managing the Data Base Environment*. Upper Saddle River, NJ: Prentice Hall PTR.

[43] PostgreSQL (2017). *JDBC Driver – GitHub*. Available at: https://github.com/pgjdbc/pgjdbc/issues/369

[44] Pereira, G., Barbosa, R., and Madeira, H. (2016). "Practical Emulation of Software Defects in Source Code," in *2016 12th European Dependable Computing Conference (EDCC)*, Gothenburg, Sweden.

Index

About the Editors

Andrea Bondavalli is a Full Professor of Computer Science at the University of Firenze. Previously he has been a researcher and a senior researcher of the Italian National Research Council, working at the CNUCE Institute in Pisa. His research activity is focused on Dependability and Resilience of critical systems and infrastructures. In particular he has been working on safety, security, fault tolerance, evaluation of attributes such as reliability, availability and performability. His scientific activities have originated more than 220 papers appeared in international Journals and Conferences. Andrea Bondavalli supports as an expert the European Commission in the selection and evaluation of project proposals and regularly consultes companies in the application field. Andrea Bondavalli led various national and European projects such as the Italian MIUR PRIN "DOTS-LCCI" and "TENACE" and the European projects ESPRIT BRA 3092 PDCS, 6362 PDCS-2, ESPRIT 20716 GUARDS, ESPRIT 27439 HIDE, IST-FP6-STREP-26979 HIDENETS, TST5-CT-2006-031413 SAFEDMI e FP7 – 216295 CA AMBER, FP7 SST-2008-234088 ALARP, the ARTEMIS-2012-1-333053 "CONCERTO", the POR CReO 2007-2013, linea di intervento 1.5.a – 1.6 "SECURE", the FP7-ICT-2013-10-610535 "AMADEOS" (coordinator), the FP7-PEOPLE-2012-IAPP-324334 "CECRIS" (Coordinator) and the PIRSES-GA-2013-612569 "DEVASSES". Andrea Bondavalli participates to (and has been chairing) the program committee in several International Conferences such as IEEE FTCS, IEEE SRDS, EDCC, IEEE HASE, IEEE ISORC, IEEE ISADS, IEEE DSN, SAFECOMP. He is the chair of the Steering Committees of IEEE SRDS and a member of the editorial board of the International Journal of Critical Computer-Based Systems. Andrea Bondavalli is a member of the IEEE, the IFIP W.G. 10.4 Working Group on "Dependable Computing and Fault-Tolerance".

Francesco Brancati took his Master degree in Computer Science at the University of Firenze in 2008 and his Ph.D. degree in Computer Science at the Resilient Computing Lab – University of Firenze in 2012. His research activity mainly focused on adaptive and safe estimation of different sources of uncertainty to improve dependability of highly dynamic systems through online monitoring analysis. During his Ph.D. he participated in the national project DOTS-LCCI (Funded by MIUR) and in the European funded project FP7-STREP-234088 ALARP) where he was mainly involved within the Architecture Design WPs, and where he worked also as Resiltech on system integration activities. Currently he works at Resiltech as Innovation Manager and SW Solution Expert, he led ResilTech participation in AMADEOS (FP7-ICT-610535) and in the CECRIS (FP7-PEOPLE-IAPP-324334) projects and he is leading the participation in the STORM project (H2020-DRS-2015-700191).